POWER TECHNOLOGY

POWER TECHNOLOGY

GEORGE E. STEPHENSON

DELMAR PUBLISHERS
COPYRIGHT © 1973
BY LITTON EDUCATIONAL PUBLISHING, INC.

LIBRARY OF CONGRESS CATALOG CARD NUMBER: 72 - 13389

PRINTED IN THE UNITED STATES OF AMERICA
PUBLISHED SIMULTANEOUSLY IN CANADA BY
DELMAR PUBLISHERS, A DIVISION OF
VAN NOSTRAND REINHOLD, LTD.

DELMAR PUBLISHERS • MOUNTAINVIEW AVENUE • ALBANY, NEW YORK 12205
A DIVISION OF LITTON EDUCATIONAL PUBLISHING, INC.

Preface

POWER TECHNOLOGY is a study of prime movers, those inventions which convert energy into work. In our technological world, we have become increasingly dependent upon prime movers and various energy sources, such as sun, water, combustion of fuels, and nuclear energy.

The objective of POWER TECHNOLOGY is to present instructional material which allows the student to gain an understanding and appreciation of energy sources and the machines which convert such energy into useful work.

Now in its third edition, POWER TECHNOLOGY continues to keep pace with the technological advances which are so rapidly changing our style of living. With more in-depth coverage and the inclusion of many recent developments, the following topics summarize the changes made in this latest edition:

● solid-state ignition

● recent quality designations and viscosity classifications for oil

● oil additives

● horizontal and vertical crankshaft engine designs

● fractional distillation of petroleum and characteristics of gasoline

● detonation and preignition

● crankcase ventilation

● spark plug characteristics and problems

● recent material on rocket engines and space exploration, including new illustrations

● new unit on pollution and its control

The author of this publication, George E. Stephenson, is a graduate of Stout State University (BS) and Colorado State University (MEd). He has also authored Small Gasoline Engines and Drawing for Product Planning. Mr. Stephenson is a past president of the Illinois Power Mechanics Association. In 1967 he received the AIAA Outstanding Teacher of the Year Award — state of Illinois. For several years, the author has served as a supervising teacher for the Western Illinois University student teaching program. Mr. Stephenson is currently a power mechanics and general industrial arts teacher in school district #205, Galesburg, Illinois.

POWER TECHNOLOGY has been designed with the small gasoline engine as a starting point. This introduction is advantageous since it lends itself to student investigation and examination in a laboratory experience type of instruction, illustrates many principles of heat engines, and is more readily accessible for school use than are larger engines.

To assist students in a more thorough study of small gasoline engines, there is a Laboratory Experiences manual. The lab experiences are designed to guide the student in the procedures of assembly and disassembly and direct his work in an orderly fashion while he investigates the workings of the small gasoline engine. Each experience is closely correlated to the appropriate unit so that the combination of theory and practice provides for a logical and unified instructional program. The procedures given in Laboratory Experiences are fundamental and apply to all makes of small gasoline engines. The 31 lab experiences were extracted from Stephenson's Small Gasoline Engines text-workbook.

Other titles in the Automotive/Power Mechanics series include:

BASIC AUTOMOTIVE SERIES (9 books)

Automobile Engine -- Basic Parts

The Brake System

The Cooling and Exhaust System

The Differential

The Fuel System

The Ignition System, Cranking and Charging Circuits

The Lubrication System

The Steering System

The Standard Transmission

AUTOMOTIVE OSCILLOSCOPE

GENERAL REPAIR TOOLS FOR AUTOMOTIVE MECHANICS

AUTOMOTIVE DRAWING INTERPRETATION

RELATED SCIENCE -- AUTOMOTIVE TRADES

PRACTICAL PROBLEMS IN MATH FOR AUTOMOTIVE TECHNICIANS

AUTOMOTIVE STEERING SYSTEMS

AUTOMOTIVE STARTING AND CHARGING SYSTEMS

AUTOMOTIVE AIR CONDITIONING

A series of full-color transparencies correlated to the POWER TECHNOLOGY units can be obtained from DCA Educational Publications, Inc., 4865 Stenton Avenue, Philadelphia, Pa. 19144. These transparencies can be employed as instructional aids to clarify and enrich the course content. A suggested procedure for the presentation of each transparency is included in the Instructor's Guide of this text.

The author and editorial staff at Delmar Publishers are interested in continually improving the quality of this instructional material. The reader is invited to submit constructive criticism and questions. Responses will be reviewed jointly by the author and source editor. Comments should be directed to the following:

Editor-in-Chief
Box 5087
Albany, New York 12205

Contents

Section I
INTRODUCTION

Section II
INTERNAL COMBUSTION ENGINES
Part I The Small Gasoline Engine

Section II
INTERNAL COMBUSTION ENGINES
Part II Other Internal Combustion Engines

Section III
EXTERNAL COMBUSTION ENGINES

Unit 1

MAN'S STRUGGLE TO HARNESS ENERGY

Prehistoric man knew power. He observed its existence in the force of the wind, the heat of the sun, in flowing water, fire, and in the wild animals that preyed on him and on one another. These forces were sometimes a blessing but very often, a real danger: heat, cold, hurricane, flood, predatory animals could threaten his existence.

When man began to think in terms of "How can I get this power to work for me?" the mechanics of power was born. Slowly through the centuries, man developed machines and devices that harnessed this energy to do his work.

Power Technology is a study of the many Energy-Converting Machines and Devices which man has developed as his technical knowledge has expanded. Nature has provided the energy sources, and man's ingenuity has provided the machines, a happy combination that is responsible for a high degree of civilization. Without his intellect and nature's help, man today would still be in his cave, hunting animals with a spear and praying that the forces of nature would be gentle.

For the general purposes of our introduction, we shall use the term "power" in its general sense, interchangeably with "force" and "energy." Later in our discussion, however, these and other terms will be defined in their true, technical descriptions.

THE SUN — MAJOR SOURCE OF POWER

Almost all sources of power can be traced to the sun. Coal was formed through the ages as plant life died and was accumulated in the swampy areas of the earth. This plant life grew through the sun's energy. When coal is burned, this energy is released — energy that has been stored for perhaps millions of years.

The story of petroleum is similar. It is believed that petroleum was formed ages ago where large amounts of plant and animal life accumulated. The petroleum fuels we burn release stored energy. Wood, of course, possesses energy and releases its energy when it is burned. Without the sun, no plants or animals could have lived.

Wind and water power are also traceable to the sun. Wind is the circulation of air caused by the sun's heat. Water power is harnessed from a flowing stream, which is filled by rainfall. The rain is caused by the sun evaporating the earth's water. In a short time the water condenses and rain falls, filling the streams.

With the exception of atomic power, all of our power sources can be traced quite directly to the sun.

EARLY ATTEMPTS TO CONTROL POWER

Surely one of man's first attempts at controlling power was to use fire for warmth and cooking.

Another early step forward was the domestication of wild animals. The taming of the horse and his use as a beast of burden provided man with new power, horse power to be exact.

Of course, the beast of burden's load was lightened with the evolution of the wheel. The wheel enabled the animal to "carry" a larger load with less effort. Again, it further increased man's power, or rate of doing work. Men used animals for land transportation for centuries and in some parts of the world animals are still the main form of transportation.

For most of the world it has only been within the last century that the beast has been taken out of harness and the engine put in its place. The oxen, donkey and horse have been replaced by the truck, tractor and automobile. No doubt our grandparents remember the "horse and buggy days" of not so long ago.

Another early force that was harnessed was wind. The use of sailboats can be traced back 5000 years or more. For centuries the exploration and conquest of the world was done with the aid of the sailing ship. Ancient traders and merchants sent their wares around the world in sailing ships. The first colonists in America were brought to our shores by sail.

DEVELOPMENT OF STEAM POWER

In 1765 James Watt of Scotland produced a successful steam engine. Although others had worked on steam engines earlier, Watt's engine was vastly more efficient. It really consisted of three parts; boiler, cylinder and piston, and condenser. When steam was let into the cylinder the piston was forced to the top of the cylinder. The steam was then shut off and the condenser opened. The condenser turned the steam back into water; when this happened a vacuum was created pulling the piston down. This cycle produced useful work and was repeated over and over. Watt's engine was first used to power water pumps in the coal mines of England.

Another type of steam engine, the turbine, was developed toward the end of the 19th century. In this engine a jet of steam directed against the turbine blade rotated the blade, producing power, much the same as a windmill. The steam turbine was used mainly for powering large ships and producing electrical power. Today the steam turbine is still widely used.

Quite naturally Watt's steam engine became an extremely efficient and refined machine through the years and was the key to rapid industrial growth. Steam engines could be built anywhere, the factory was no longer tied to the river and water power. The steam engine provided the power for the newly developing factories of the era. Workers and craftsmen were brought from their small shops into the more efficient factories. Steam-powered machinery increased the productivity of workers. England attempted a monopoly on the use of steam engines, but in 1789 Samuel Slater came to America with memorized plans. By 1807, fifteen steam powered cotton mills were operating successfully in America.

On August 18, 1807, Robert Fulton boarded his new ship, the Clermont, fired the stokers of a steam engine and, in the midst of flying sparks and thick smoke, the first successful steamship voyage was made. Fulton traveled from New York to Albany in a record time of 32 hours.

Steam engines were soon used to power railroads in the U. S. Although the race between Peter Cooper's Tom Thumb and a horse-drawn train in 1830 failed due to mechanical difficulties, the steam engine DeWitt Clinton followed by making a successful run from Albany, New York to nearby Schenectady.

THE DEVELOPMENT OF THE INTERNAL COMBUSTION ENGINE

Steam power was dominant as a power source until the advent of the internal combustion engine. The internal combustion engine has taken over many steam engine jobs and has found a multitude of new jobs on its own. The smaller, internal combustion engine was ideally suited for land vehicles and the development of automobiles, trucks, tractors, etc.

One of the earliest attempts to produce an internal combustion engine was made by Christian Huygens late in the 17th century. In his engine a piston was forced down a cylinder by an explosion of gunpowder. Many other persons continued working through the years to produce a successful engine in which fuel could be burned directly, right inside the engine itself.

The first practical internal combustion engines were produced in 1878, less than 100 years ago. These engines were made by the Otto and Langan firm at Deutz, Germany. They burned the vapor from oil for fuel and, though successful, they were crude and cumbersome by present standards. The engines weighed about 1,110 pounds per horsepower. Today, engines are manufactured that weigh as little as three pounds per horsepower.

Smaller internal combustion engines such as the "Small Gasoline Engine" also have earned their place in advancing our civilization. They, too, have been helping man to do his work for many years. And in recent years small engines have found a tremendous popularity both for work and recreation.

CURRENT POWER DEVELOPMENTS

The world of power and engines is wide. New applications are found for current engines and new engines are being constantly developed. Power sources are also under continual research and development. Electrical power, though not new, is a major source of power. Diesel engines, gas turbine engines, jet engines, rocket engines, and atomic engines are all topics for today's student in power mechanics.

GENERAL STUDY QUESTIONS

1. What is the original source of nearly all power? Explain.

2. What sources of power were used before the steam engine?

3. In what ways did the development of the steam engine affect society?

4. Who is credited with producing the first practical internal combustion engine? How long ago?

5. Compare the weight per horsepower of early engines with the weight per horsepower of present day engines.

6. How has the internal combustion engine affected our modes of travel, economy, and way of life?

7. List several uses of small gasoline engines, both for work and pleasure.

CLASS DISCUSSION TOPICS

● Discuss how machines evolved slowly through the centuries.

● Discuss the rapid technological developments of the 19th and 20th centuries.

● Discuss the steam engine and the Industrial Revolution.

● List the common types of heat engines.

● Discuss the occupations in which small gasoline engines are regularly used.

● Discuss how friends, relatives and others use small gasoline engines occasionally during the year.

Unit 2

WORK, ENERGY, POWER

In Unit 1 we briefly reviewed man's attempts to control sources of power from his earliest awareness of such sources to the present day. Before getting into a detailed study of the elements of small gasoline engines, however, it is necessary for us to review some of the terms which are basic to an understanding of mechanical power. What is power? What is energy? What is work? Are these terms all the same and, if not, how do they differ? It is the purpose of this unit to define and explain these terms, so that as they are used in later units their meaning will be clear.

WORK

Work is a scientific term as well as an everyday term. To the man on the street, work means engaging in an occupation. One person may work hard at an office while another person may work hard carrying bricks on a construction job. Both men may come home exhausted but, scientifically speaking, only the man carrying bricks has done much work.

Work involves moving things, and more than that, moving things by applying a force. Work, then, is applying a force to cause motion or in other words, motion caused by applying a force.

Measurement of Work

The common unit for measuring work is the foot-pound. Raising one pound, one foot is one foot-pound of work. A man carrying 50 pounds of bricks up a 10-foot ladder does 500 foot-pounds of work. Work is, therefore, applying a force through a distance.

Work   = Force   ×   Distance

(Ft.-lbs.) = (Pound) ×   (Feet)

As a simple example, calculate the work involved in lifting a 20-pound weight 5 feet.

Work =   Force × Distance

Work =   20 lbs. × 5 ft. = 100 ft.-lbs.

By scientific definition, the man who struggles to move a boulder but fails to budge it is not doing work because there is no "distance moved". Of course all engines can do work because they are capable of applying force to move machines, loads, implements, and so forth, through distances.

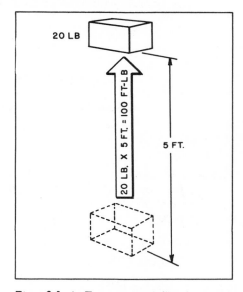

Fig. 2-1 A Twenty-pound Weight Lifted Five Feet in the Air.

ENERGY

Energy is the ability to do work. The energy stored in our bodies, for instance, gives each of us a potential to do work. A person can work for a long period of time on his stored energy before he needs "refueling" with food. The human body is a good example of one major energy form: potential energy.

Potential Energy

Potential energy is scientifically stated as the energy a body has due to its position, its condition, or its chemical state: position, water at the top of a waterfall; condition, a tightly wound watch spring; chemical state, fuels such as gasoline, coal, etc., or foodstuffs to be burned in the body.

5

Stored or potential energy is measured in the same way as work, in foot-pounds. How much potential energy is there when a 20-pound weight is lifted 5 feet?

Potential Energy = Force × Distance

Potential Energy = (weight) × (height)

Potential Energy = 20 lbs. × 5 ft.

Potential Energy = 100 ft.-lbs.

With the weight at the five-foot height, there is 100 foot-pounds of potential energy.

Kinetic Energy

Kinetic energy is the energy of motion, the energy of a thrown ball, the energy of water falling over a dam, the energy of a speeding automobile. Kinetic energy is, in effect, "released potential energy".

Consider a thrown ball, the work of throwing it, and what becomes of it. If a 3/4-pound ball is thrown by applying a force of 8 pounds through 6 feet, how much work is done?

Work = Force × Distance

Work = 8 lbs. × 6 ft.

Work = 48 ft.-lbs.

What becomes of this energy? It exists as energy of motion of the ball, kinetic energy. The ball in flight has 48 ft.-lbs. of energy which it will deliver to whatever it hits.

Turning back to the example of potential energy, when the 20-pound weight which rests 5 feet above the floor is allowed to fall, its potential energy of 100 ft.-lbs. becomes 100 ft.-lbs. of kinetic energy, energy of motion.

Energy can change form but it cannot be destroyed. This is the Law of the Conservation of Energy. Energy can take the form of light, sound, heat, motion, and electricity. Consider the potential chemical energy of gasoline in an engine. Upon ignition, its energy is converted mainly into heat energy. The heat energy is used to push the pistons down and now can be seen as the kinetic energy of the rotating crankshaft.

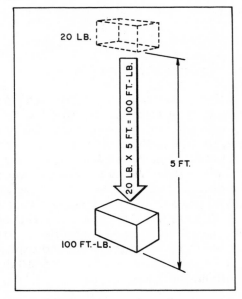

Fig. 2-2 A Twenty-pound Weight Dropped to the Floor from a Height of Five Feet Delivers 100 ft.-lbs. of Kinetic Energy.

Again refer back to the 20-pound weight poised at a 5-foot height. When the weight falls to the floor its energy is not lost, but, rather, it is transformed into other types of energy. The floor would become slightly warmer at the point of impact due to the creation of heat energy. There would be an audible sound at the moment of impact, an indication of the presence of sound energy. Of course, the floor would give slightly, thus absorbing some energy.

The Efficiency of Machines that transform energy is not 100 percent. That is to say, some used energy does not perform a useful purpose. It literally may be wasted in moving the machine parts to overcome friction inherent in all machines. The efficiency of a machine is the ratio of the work done by the machine to the work put into it.

Stated as a formula:

$$\text{Efficiency} = \frac{\text{Output}}{\text{Input}} \times 100$$

or

$$\text{Efficiency} = \frac{\text{Input - Losses}}{\text{Input}} \times 100$$

or

$$\text{Efficiency} = \frac{\text{Output}}{\text{Output}} + \text{Losses} \times 100$$

The figure "100" in these formulas is a way of converting the fraction to a percentage.

POWER

In everyday language people use the word "power" to mean a variety of things: "political power", "financial power", etc. However, in the language of scientists and engineers, "power" refers to how fast work is done, or how fast energy is transferred. "Power is the rate of doing work" and "Power is the rate of energy conversion" are two useful definitions of the term.

How fast a machine can work is an important consideration for engineers, and for consumers too. For example, when gasoline is purchased the person buys potential energy. The "size" of the engine will determine how fast this energy can be converted into power. A 10-horsepower engine can work only half as fast as a 20-horsepower engine.

Measurement of Power

Power is measured in foot-pounds per second or in foot-pounds per minute. Referring again to the 20-pound weight that was lifted 5 feet, how much power was required to lift the weight if it were done in 5 seconds?

Work = Force × Distance

Work = 20 lbs. × 5 ft. = 100 ft.-lbs.

$$\text{Power} = \frac{\text{Work}}{\text{Time}}$$

$$\text{Power} = \frac{100 \text{ ft.-lbs.}}{5 \text{ seconds}}$$

Power = 20 ft.-lbs./second

As another example: if an elevator lifts 3500 pounds a distance of 40 feet and it takes 25 seconds to do it, what is the rate of doing work?

Work = Force × Distance

Work = 3500 lbs. × 40 ft. = 140,000 ft.-lbs.

$$\text{Power} = \frac{\text{Work}}{\text{Time}}$$

$$\text{Power} = \frac{140,000 \text{ ft.-lbs.}}{25 \text{ sec.}} = 5600 \text{ ft.-lbs./sec.}$$

Power can also be expressed in foot-pounds per minute. If a pump needs 10 minutes to lift 5000 pounds of water 60 feet, it is doing 300,000 ft.-lbs. of work in 10 minutes, which is a rate of 30,000 foot-pounds per minute.

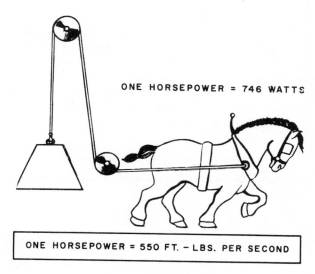

ONE HORSEPOWER = 746 WATTS

ONE HORSEPOWER = 550 FT. – LBS. PER SECOND

Fig. 2-3 Horsepower.

Horsepower

The power of most machinery is measured in horsepower. The unit originated many years ago when James Watt, attempting to sell his new steam engines, had to rate his engines in comparison with the horses they were to replace. He found that an average horse, working at a steady rate, could do 550 foot-pounds of work per second. This rate is the definition of one horsepower.

The formula for horsepower is:

$$\text{Horsepower} = \frac{\text{Work}}{\text{Time (in seconds)} \times 550}$$

If Time in the formula above is expressed in minutes, it is multiplied by 550 × 60 (seconds) or 33,000. The formula, then, may be expressed as:

$$\text{Horsepower} = \frac{\text{Work}}{\text{Time (in minutes)} \times 33,000}$$

What horsepower motor would the elevator previously referred to have?

$$\text{Horsepower} = \frac{\text{Work}}{\text{Time (in seconds)} \times 550}$$

$$\text{Horsepower} = \frac{140,000 \text{ ft.-lbs.}}{25 \text{ sec.} \times 550}$$

Horsepower = 10 +

SUMMARY

We have now discussed those terms which are basic to an understanding of the applications of mechanical power which will follow in later units. The measurements which these terms involve have been shown as formulas. To review, the following formulas are a means of expressing each of the terms we have covered:

a. $WORK = FORCE \times DISTANCE$

b. $EFFICIENCY = \dfrac{OUTPUT}{INPUT} \times 100$

c. $POWER = \dfrac{WORK}{TIME} = \dfrac{FORCE \times DISTANCE}{TIME}$

d. $HORSEPOWER = \dfrac{WORK}{TIME \text{ (in sec.)} \times 550}$

$= \dfrac{WORK}{TIME \text{ (in min.)} \times 33000}$

GENERAL STUDY QUESTIONS

1. Give a definition of work.

2. What is the unit of measurement for work?

3. A 100-pound boy climbs a 16-foot stairway. How much work has he done?

4. Define energy.

5. Explain the difference between kinetic and potential energy.

6. Explain the conservation of energy. What are some common forms of energy?

7. Explain why a machine cannot be 100 percent efficient.

8. Define power.

9. What is the unit of measurement for power?

10. How much power is needed to lift a 100-pound bag of cement onto a 3-foot truck bed in 2 seconds?

11. What is the formula for horsepower?

12. How much horsepower does an engine on a grain elevator deliver if it can load 705 pounds of corn into a 25-foot storage bin in 55 seconds?

CLASS DISCUSSION TOPICS

● Have a student lift a given weight a definite height and calculate the work.

● Discuss the difference between "scientific work" and other kinds of work.

● Discuss and list many potential and kinetic energy sources.

● Discuss the origin of the term horsepower.

CLASS DEMONSTRATION TOPICS

▸ Demonstrate how the horsepower developed by an average student might be calculated.

▸ Demonstrate the transformation of energy by striking and burning a match.

▸ Demonstrate potential and kinetic energy sources.

▸ Demonstrate the conservation of energy.

▸ Demonstrate losses due to friction and how they affect the efficiency of a machine.

Part I The Small Gasoline Engine

Unit 3
CONSTRUCTION OF THE SMALL GASOLINE ENGINE

The internal combustion engine is classified as a <u>heat</u> engine; in other words, its power is produced by <u>burning</u> a fuel. The power stored in the fuel is released when it is burned. The word "internal" means that the fuel is burned inside the engine itself. Our most common fuel is gasoline. Of course, if gasoline is to burn inside the engine there must be oxygen present to support the combustion. Therefore, the fuel actually needs to be a mixture of gasoline and air. When ignited, a fuel mixture of gasoline and air burns ferociously; it almost explodes. The engine is designed to harness this power.

The engine block is fitted with a cylindrical shaft. This is the <u>cylinder</u>. A plate fits over the head of the block, sealing off the top of the cylinder, the <u>cylinder head</u>. The cylinder contains a <u>piston</u>, a cylindrical fitting which fits the cylinder exactly. The piston is free to slide up and down within the cylinder. The fuel mixture is brought into the cylinder; then the piston moves up and compresses the fuel into a small space called the combustion chamber. When the fuel is ignited and burns, tremendous pressure builds up. This pressure forces the piston back down the cylinder; thus the untamed energy of combustion is harnessed to become useful mechanical energy. The basic motion within the engine is that of the piston sliding up and down the cylinder, a reciprocating motion.

There are still many problems, however. How can the up-and-down motion of the piston be converted into useful rotary motion? How can exhaust gases be removed? How can new fuel mixture be brought into the combustion chamber? Let us look at the engine's basic parts.

BASIC ENGINE PARTS

The internal combustion engine parts discussed on the next few pages are those of the four-stroke cycle gasoline engine, our most common type. The most essential of these parts are briefly listed below.

- Cylinder: hollow; stationary; piston moves up and down within cylinder.
- Piston: fits snugly into cylinder but still can be moved up and down.
- Crankshaft: converts reciprocating motion into more useful rotary motion.
- Connecting rod: connects piston and crankshaft.
- Valves: "doors" for admitting fuel mixture and releasing exhaust.
- Crankcase: the body of the engine, it contains most of engine's moving parts.

CYLINDER

The cylinder is a finely machined, high quality cast iron part in which the piston slides up and down. Even though an engine may be basically made from aluminum, the cylinder itself will usually be made of cast iron. Normally the aluminum body of the engine is cast around the cylinder and the two are inseparable.

However, on some engines the cylinder section can be removed. This is particularly true of very large engines. In engine specifications the diameter of the cylinder is an important measurement. It is referred to as the "bore" of the engine.

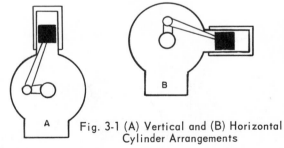

Fig. 3-1 (A) Vertical and (B) Horizontal Cylinder Arrangements

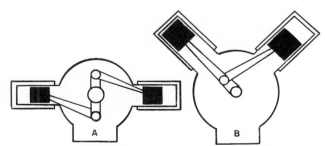

Fig. 3-2 (A) Opposed and (B) V-type Cylinder Arrangements

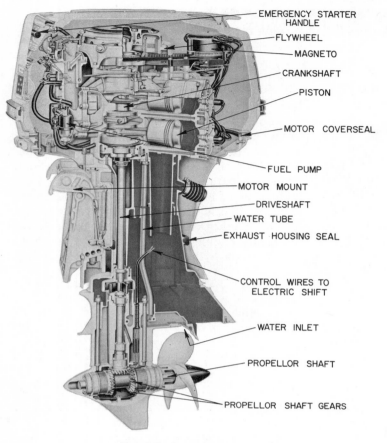

Fig. 3-3 Cross-section of a Typical
Johnson Outboard Motor.

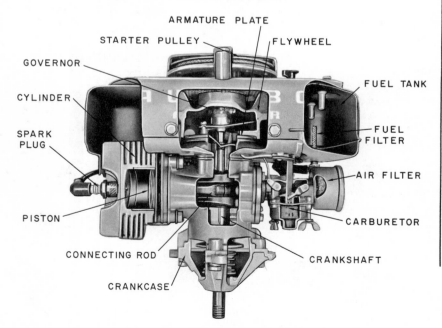

Fig. 3-4 Cross-section of a Typical Lawn-Boy Engine.

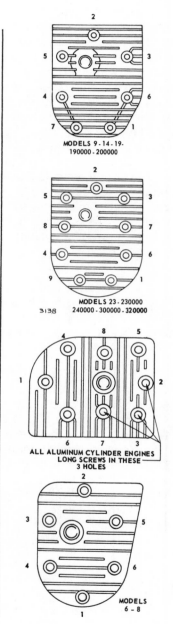

MODELS 9 - 14 - 19 -
190000 - 200000

3138   MODELS 23 - 230000
240000 - 300000 - 320000

ALL ALUMINUM CYLINDER ENGINES
LONG SCREWS IN THESE
3 HOLES

MODELS
6 - 8

Fig. 3-5 A Typical Cylinder Head.
Note the Sequence for Tightening
the Head Bolts.

CYLINDER HEAD

The cylinder head forms the top of the combustion chamber and it is bolted tightly to the cylinder. It is important to tighten all cylinder head bolts with an even pressure and in their correct order so that uneven stresses will not set up in the cylinder walls. A gasket between the two metal surfaces makes an airtight seal. The cylinder head also contains the spark plug.

Fig. 3-7 Pistons Cast of Aluminum.

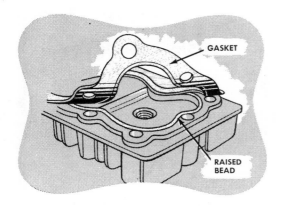

Fig. 3-6 The Cylinder Head Gasket Provides an Airtight Seal.

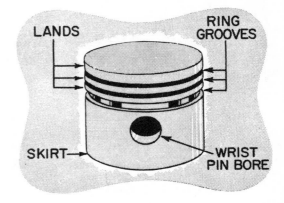

Fig. 3-8 The Nomenclature of the Piston

PISTON

The piston slides up and down in the cylinder. It is the only part of the combustion chamber that can move when the pressure of the rapidly burning fuel mixture is applied. The piston may be made of cast iron, steel, or aluminum; aluminum is commonly used for small engines because of its light weight and its ability to conduct heat away rapidly.

There must be clearance between the cylinder wall and the piston to prevent excessive wear. The piston is .003 to .004 of an inch smaller than the cylinder. To check this clearance, a special feeler gage is necessary.

The piston top, or "crown" may be flat, convex, concave, or any of many other shapes. Manufacturers select the shape that will cause turbulence of the fuel mixture in the combustion chamber and that will promote smooth burning to the fuel mixture. The piston has machined grooves near the top to accommodate the piston rings.

The distance the piston moves up and down is called the "stroke" of the engine. The uppermost point of the piston's travel is called its top dead center position (TDC). The lowest point of the piston's travel is called its bottom dead center (BDC). Therefore, "stroke" is the distance between TDC and BDC.

PISTON RINGS

Piston rings provide a tight seal between the piston and the cylinder wall. Without the piston rings much of the force of combustion would escape between the piston and cylinder into the crankcase. Primarily the piston rings act as a "power seal", ensuring that a maximum amount of the combustion power is used in forcing the piston down the cylinder. Of course, by having the piston rings between the piston and cylinder wall there is a small area of metal sliding against the cylinder wall. Therefore, the piston rings reduce friction and the accompanying heat and wear that are caused by friction. Another function of the piston ring is to control the lubrication of the cylinder wall.

The job of the piston rings might appear rather simple but the piston rings may have to work under several harsh conditions such as, (1) distorted cylinder walls, due to improper tightening of head bolts, (2) cylinder out of round, (3) worn or scored cylinder walls, (4) worn piston, and (5) conditions of expansion due to intense heat. The important job of providing a power seal can become difficult if the engine is worn or has been abused.

Piston rings are made of cast iron or steel and are finely machined. Sometimes their surfaces are plated with other metals to improve their action. There are many designs of piston rings and some rings, especially oil rings, consist of more than one piece.

Two serious problems, (1) blow-by and (2) oil pumping, can be caused by bad piston rings, bad piston, warped cylinder, distortion, and/or scoring. In blow-by, the pressures of combustion are great enough to break the oil seal provided by the piston rings. When this happens, the gases of combustion force their way into the crankcase. The result is a loss in engine power and contamination of the crankcase.

Piston rings that are not fitted properly may start pumping oil. This pumping action will lead to fouling in the combustion chamber, excessive oil consumption, and poor combustion characteristics. The rings have a side clearance which can cause a pumping action as the piston moves up and down.

A piston will usually have three or four piston rings. The top piston ring is a compression ring which exerts a pressure of 8-12 pounds on the cylinder wall. The ring has a clearance in its groove; it can move slightly up and down and expand slightly in and out. The second ring from the top is also a compression ring. These two rings form the power seal.

The third ring down, and fourth if one is present, is an oil control ring. The job of this piston ring is to control the lubrication of the cylinder wall. These rings spread the correct amount of oil on the cylinder wall, scraping the excess from the wall and returning it to the crankcase. These rings are slotted and grooves are cut into the piston behind the ring. This enables much of the excess oil to be scraped

"through" the piston and it then drips back into the crankcase. It should be noted that compression rings have a secondary function of oil control.

Fig. 3-9 Typical Compression Ring.

Fig. 3-10 Typical Oil Control Ring.

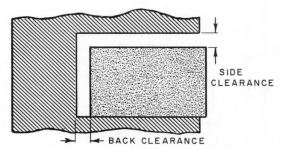

Fig. 3-11 Cross-section Showing a Piston Ring in its Groove. Notice Back Clearance and Side Clearance.

Two clearances or tolerances that are important in piston rings are, (1) end gap and (2) side clearance. The end gap is the space between the ends of the piston ring, measured when the ring is in the cylinder. The gap must be large enough to allow for expansion due to heat but not so large that power loss due to blow-by will result. Often .004" is allowed for each inch of piston diameter (top ring), .003" is allowed for rings under the top ring. Since the second and third rings are exposed to less heat their allowance can be smaller.

Side clearance or ring groove clearance is also provided for heat expansion. Often .0025" is allowed for the top ring and .003" is allowed on the second and third rings.

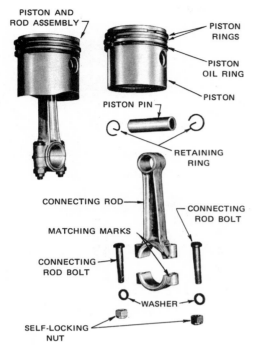

Fig. 3-12 A Typical Piston, Piston Pin, Connecting Rod, and Associated Parts.

PISTON PIN OR WRIST PIN

The piston pin is a precision ground steel pin that connects the piston and the connecting rod. This pin may be solid or hollow; the hollow piston pin has the advantage of being lighter in weight. Because piston pins are subjected to heavy shocks when combustion takes place, high tensile strength steel is used. To keep noise at a minimum, the piston pin fits to a very close tolerance in the connecting rod. The piston pin does not rotate. It has a rocking motion similar to the action of a man's wrist as he holds a bar tightly and swings it back and forth: hence the name "wrist pin." Heavy shocks, close tolerances, and the rocking motion make the lubrication of this part difficult. In most engines, lubricating oil is squirted or splashed on it.

CONNECTING ROD

The connecting rod connects the piston (with the aid of the piston pin) and the crankshaft. Many small engines have cast aluminum connecting rods; larger engines may have steel connecting rods. Generally the cross section of the connecting rod is an "I" beam shape. The

lower end of the connecting rod is fitted with a cap. Both the lower end and the cap are accurately machined to form a perfect circle. This cap is bolted to the rod, encircling the connecting rod bearing surface on the crankshaft. Many connecting rods are fitted with replaceable bearing surfaces, especially on heavy-duty or more expensive engines.

CRANKSHAFT

The crankshaft is the vital part that converts the reciprocating motion of the piston into rotary motion. One end of the crankshaft has a provision for power takeoff and the other is machined to accept the flywheel.

Crankshafts are carefully forged and machined steel. Because they must accept a great amount of force, their bearing surfaces are large, and because they revolve at high speeds, they must be well balanced.

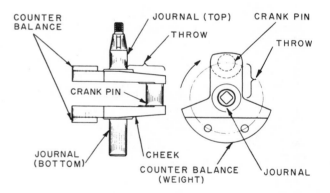

Fig. 3-13 Single Throw Crankshaft: One Crankpin for One Cylinder.

Engines designed with vertical crankshafts are excellent for applications such as rotary power lawnmowers where the blade is bolted directly to the end of the crankshaft. Most wheeled applications such as motor bikes and go-karts use horizontal crankshaft engines which enable the power to be delivered to an axle through belts or chains. Multi-position engines such as chain saws are another variation. The crankshafts themselves are much the same but other changes in engine design are necessary to meet the requirements of crankshaft position. An engine may also have clockwise or counterclockwise crankshaft rotation.

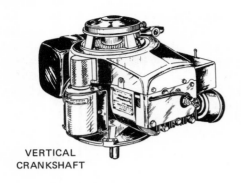

VERTICAL
CRANKSHAFT

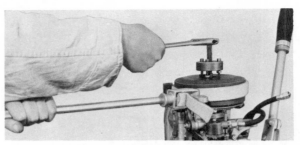

Fig. 3-15A Pulling the Fly Wheel of an Outboard Engine.

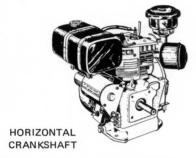

HORIZONTAL
CRANKSHAFT

Fig. 3-14 Two Types of Small Engine Design.

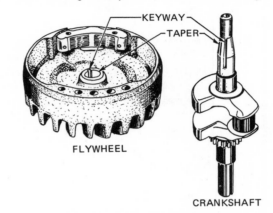

KEYWAY
TAPER
FLYWHEEL
CRANKSHAFT

Fig. 3-15B A Key "Locks" the Fly Wheel and Crank
Shaft Together.

CRANKCASE

The crankcase is the "body" of the engine. It houses the crankshaft and has bearing surfaces on which the crankshaft revolves. Many other parts are located within the crankcase: connecting rod, cam gear and camshaft, lubrication mechanism. The cylinder and bottom of the piston are exposed to the crankcase. On four-cycle engines a reservoir of oil is found within the crankcase. This lubricating oil is splashed, pumped, or squirted onto all the moving parts located in the crankcase.

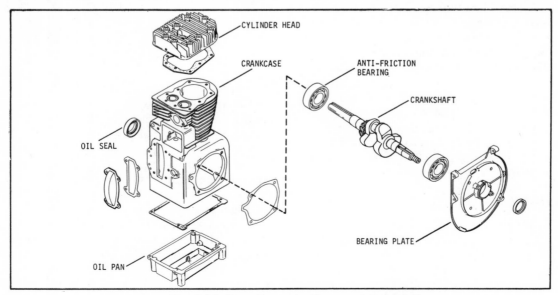

CYLINDER HEAD

CRANKCASE

ANTI-FRICTION
BEARING

CRANKSHAFT

OIL SEAL

BEARING PLATE

OIL PAN

Fig. 3-16 Major Parts of a Horizontal Crankshaft.

FLYWHEEL

The flywheel is mounted on the tapered end of the crankshaft. A key keeps the two parts solidly together and they revolve as one part. The flywheel is found on all small engines. It is relatively heavy and helps to smooth out the operation of the engine. That is, after the force of combustion has pushed the piston down, the momentum of the flywheel helps to move the piston back up the cylinder. This tends to minimize or eliminate sudden jolts of power during combustion. The more cylinders the engine has, the less important this "smoothing out" action becomes.

The flywheel may also be a part of the engine's cooling system by having air vanes to scoop air as the flywheel revolves. This air is channeled across the hot engine to carry away heat.

Most flywheels on small gasoline engines are part of the ignition systems, in that they have permanent magnets mounted in them. These permanent magnets are an essential part of the magneto ignition system.

VALVES

The most common valves used today are called poppet valves. With these valves, exhaust gases can be removed from the combustion chamber and new fuel mixture can be brought in. To accomplish this, the valve actually pops open and snaps closed. In the normal four-cycle combustion chamber there are two valves, one to let the exhaust out (Exhaust) and one to allow the fuel mixture to come in (Intake).

Quite possibly, the two valves will look almost identical, but they are different. The exhaust valve must be designed to take an extreme amount of punishment and still function perfectly. Not only is it subjected to the normal heat of combustion (about 4500° F), but when it opens, the hot exhaust gases rush by it and when the valve returns to its seat, a small area touches, making it difficult for the valve's heat to be conducted away. This rugged valve opens and closes 1800 times a minute if the engine is operating at 3600 r.p.m.

Exhaust valves are made from special heat-resisting alloys and are hollow, being filled with metallic sodium. This metallic sodium melts at about the boiling point of water and helps to rapidly conduct the heat from the valve head down the valve stem. Some exhaust valves, however, are made from solid steel.

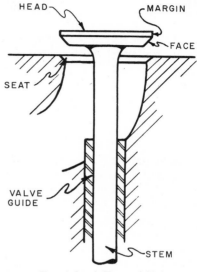

Fig. 3-17 A Typical Valve.

Intake valves operate in the same manner as exhaust valves except they are not subjected to the extreme heat conditions. Each time they open new fuel mixture rushes by, helping to cool the intake valve.

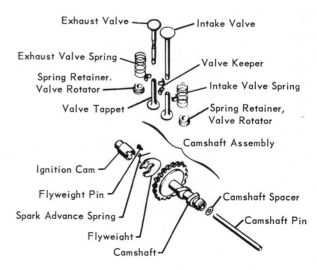

Fig. 3-18 Complete Valve Train and Camshaft Assembly.

Operation of the valve requires the help of associated parts or "valve train." The valve

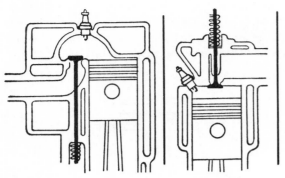

Fig. 3-19 L-head Valve Arrangement

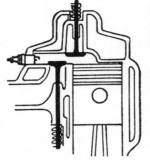

Fig. 3-20 I-head Valve Arrangement.

Fig. 3-21 F-head Valve Arrangement.

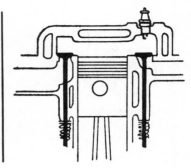

Fig. 3-22 T-head Valve Arrangement.

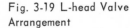

train consists of the valve (intake or exhaust), valve spring, tappet or valve lifter, and cam. The valve spring closes the valve, holding it tightly against its seat. The tappet rides on the cam and there is a small clearance between the tappet and the valve stem; some tappet clearances are adjustable especially on larger and more expensive engines. The cam or lobe pushes up on the tappet; the tappet pushes up on the valve stem; and the valve opens when the high spot on the cam is reached. Of course the opening and closing of the valves must be carefully timed in order to get the exhaust out at the correct time and the new fuel mixture in at the correct time.

VALVE ARRANGEMENT

Engines may be designed and built with one of several valve arrangements.

- The L-head: both valves open up on one side of the cylinder. This is the most common arrangement in small gasoline engines

- The I-head: both valves open down over the cylinder.

- The F-head: one valve opens up and one valve opens down; both are on the same side of the cylinder.

- The T-head: one valve is on one side of the cylinder and the other valve is on the opposite side; both valves open up.

CAMSHAFT

The camshaft and its associated parts control the opening and closing of the valves.

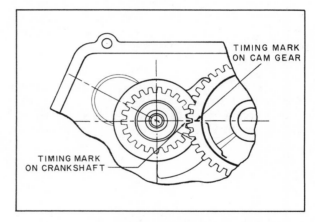

Fig. 3-23 Aligning Timing Marks on the Crankshaft Gear and Camshaft Gear. Notice the Larger Camshaft Gear Revolves at One-half Crankshaft Speed.

The cams on this shaft are eggshaped. As they revolve, their noses cause the valves to open and close. A tappet rides on each cam and when the nose of the cam comes under the tappet the valve is pushed open, admitting new fuel mixture or allowing exhaust to escape, depending on which valve is opened.

The camshaft is driven by the crankshaft, and on four-cycle engines the camshaft operates at one-half the crankshaft speed. (If the engine is operating at 3400 r.p.m., the camshaft will be revolving at 1700 r.p.m.) The crankshaft and camshaft must be in perfect synchronization, so most engines have timing marks on the gears to insure the correct reassembly of the gears.

FOUR-STROKE CYCLE OPERATION

All of these parts and many others must be assembled to operate as a smooth functioning, powerful team; the gasoline engine. The most common method of engine operation is the four-stroke cycle. Most small gasoline engines operate on this principle as do their larger brothers, automobile engines.

Four-stroke cycle means that it takes four strokes of the piston to complete the operating cycle of the engine. The piston goes down the cylinder, up, down, and back up again to complete the cycle. This takes two revolutions of the crankshaft; each stroke is one-half revolution.

When the piston travels down the cylinder, this is Intake; fuel mixture enters the combustion chamber. Reaching the bottom of its stroke, the piston comes up on Compression and the fuel mixture is pressed into a small space at the top of the combustion chamber. Power comes when the fuel mixture is ignited, pushing the piston back down. The piston moves up once more pushing the exhaust gases out of the combustion chamber. This is Exhaust. These four strokes complete the engine's cycle and require two revolutions of the crankshaft. As soon as one cycle is complete, another begins. If an engine were operating at a speed of 3600 r.p.m., there would be 1800 cycles each minute.

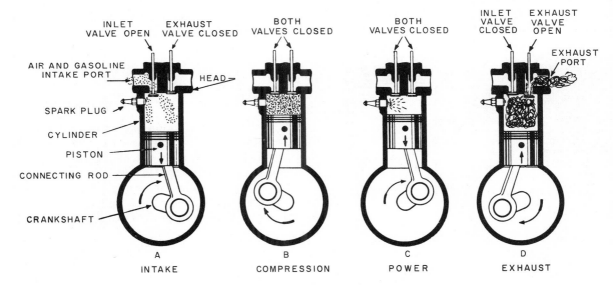

Fig. 3-24 The Four-stroke Cycle.

It would be a good idea to take a closer look at the four strokes of the four-stroke cycle.

INTAKE STROKE

If the fuel mixture is to enter the combustion chamber, the intake valve must be open, the exhaust valve must be closed. With the intake valve open and the piston traveling down the cylinder, the fuel mixture rushes in easily.

All around us and pushing down on everything is an atmospheric pressure of 14.7 lbs. per sq. in. at sea level. When the piston is at the top of its stroke there is normal atmospheric pressure in the combustion chamber, but when the piston travels down the cylinder there is more space for the same amount of air and the air pressure is reduced; a partial vacuum is created. The intake valve opens, the normal atmospheric pressure rushes in to equalize this lower air pressure in the combustion chamber. The fuel mixture is actually pushed into the combustion chamber, even though we think of it as being sucked in by the partial vacuum.

COMPRESSION STROKE

When the piston reaches the bottom of its stroke and the cylinder is filled with fuel mixture, the intake valve closes and the exhaust valve remains closed. The piston now travels up the cylinder compressing the fuel mixture into a smaller and smaller space. When the fuel mixture is compressed into this small space it can be ignited more easily and it will expand very rapidly.

The term "Compression Ratio" applies to this stroke. If the piston compresses the fuel mixture into one-sixth of the original space the compression ratio is 6:1. An engine whose piston compresses the fuel mixture into one-eighth the original space has a compression ratio of 8:1. The higher the compression, the more powerful the force of combustion will be.

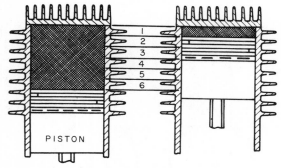

Fig. 3-25 An Engine with a Six to One Compression Ratio.

POWER STROKE

When the piston reaches the top of its stroke, and the fuel mixture is compressed, a spark jumps across the spark plug igniting the fuel mixture. The fuel mixture burns very rapidly, the next thing to an explosion, and the burning, expanding gases exert a great pressure in the combustion chamber. The combustion pressure is felt in all directions, but only the piston is free to react. This it does by being pushed rapidly down the cylinder. This is the power stroke: rapidly burning fuel creates a pressure that pushes the piston down, turning the crankshaft and producing usable rotary motion.

Of course both intake and exhaust valves must be closed for this stroke. The force of combustion must not leak out the valves. Also, this force should not "blow-by" the piston rings.

Correctly fitting piston rings prevent this. All gaskets, namely the cylinder head gasket and spark plug gasket, must be airtight. The power must be transmitted to the top of the piston and not lost elsewhere.

EXHAUST STROKE

When the piston reaches the bottom of its power stroke the momentum of the flywheel and crankshaft bring the piston back up the cylinder. This is the exhaust stroke. Now the exhaust valve opens, the intake valve remains closed, and the piston pushes the burned exhaust gases out of the cylinder and combustion chamber.

The four-stroke cycle engine is an efficient and entirely successful producer of power.

TWO-STROKE CYCLE ENGINE

Another common operating principle for gasoline engines is the two-stroke cycle. This engine is also entirely successful and in common use today, especially for outboard motors, chain saws, and many lawnmowers. The two-stroke cycle engine is constructed somewhat differently from the four-stroke cycle engine.

Two-stroke cycle means that it takes two strokes of the piston to complete the operating cycle of the engine. The piston goes down the cylinder and back up again and the cycle is completed. This takes only one revolution of the crankshaft.

With the piston at the top of its stroke, the fuel mixture tightly compressed in the combustion chamber, the engine is ready for its first stroke, the Power stroke. A spark ignites the fuel mixture and the great pressure pushes the piston rapidly down the cylinder. As the piston nears the bottom of its stroke, the exhaust ports begin to uncover and the exhaust gases, which are still hot and under pressure, start to rush through the ports and out of the engine. Just after the exhaust ports begin to uncover, the piston uncovers the intake ports. Fuel mixture rushes into the cylinder, being deflected to the top of the cylinder first and

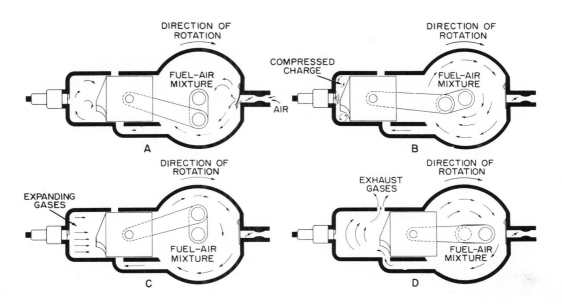

Fig. 3-26 Two-cycle Engine Operation.

finally scavenging out the last bit of exhaust. By this time the piston has reached the bottom of its stroke and is on its way back up the cylinder; the ports are sealed off, trapping the fuel mixture. Therefore, this stroke is Compression, the piston pushing the fuel mixture into a smaller and smaller space. The two strokes, then are Power and Compression, with Intake and Exhaust taking place between the two.

One question remains to be answered: why does the fuel mixture rush through the intake ports into the combustion chamber? First of all, not because there is a partial vacuum in the combustion chamber as in the four-stroke cycle engine. With the exhaust ports open, the combustion chamber soon would be at atmospheric pressure. The answer lies in the crankcase and reed valves. As the piston moves up the cylinder, a low pressure, or partial vacuum, is created in the crankcase; there is a larger space for the same amount of air. The greater atmospheric pressure outside "sees" this low pressure and rushes through the carburetor, pushing open the springy reed valves, filling the crankcase with fuel mixture. When the pressure in the crankcase and the atmospheric pressure are just about equal, the leaf valves spring shut.

As the piston moves back down the cylinder on its power stroke, the fuel mixture, trapped in the crankcase is put under a slight pressure; the same amount of mixture being pushed into a smaller space. The leaf valve can open in only the opposite direction. When the piston nears the bottom of its stroke, the exhaust ports uncover allowing the exhaust to escape. The intake ports open slightly after the exhaust ports and the pressure in the crankcase pushes through the intake ports and into the cylinder, filling it with new fuel mixture.

This cycle would then be repeated. If the engine operates at 4000 r.p.m. the cycle will be completed 4000 times each minute. There is a power stroke for each revolution of the crankshaft.

CRANKCASE

The crankcase of a two-stroke cycle engine is designed to be just as small in volume as possible. The smaller the volume, the greater the pressure created as the piston comes back down the cylinder. The greater the crankcase pressure, the more efficient the transfer of fuel from the crankcase through the transfer ports into the cylinder.

INTAKE AND EXHAUST PORTS

Intake and exhaust ports are holes drilled in the cylinder wall to allow exhaust gases to escape and new fuel mixture to enter. They are located near the bottom of the cylinder and are covered and uncovered by the piston. By using ports, many parts seen in the four-cycle engine are eliminated: valves, tappets, valve springs, cam gear, camshaft. The remaining major moving parts are the piston, connecting rod, and crankshaft.

PISTON

The piston still serves the same function in the cylinder, but for two-cycle engines the top is designed differently. On the intake side there is a sharp deflection that sends the incoming fuel mixture to the top of the cylinder. On the exhaust side there is a gentle slope so that exhaust gases have a clear path to escape. The loop scavenged two-cycle engine does, however, have a flat top piston; it will be discussed in a later section.

REED OR LEAF VALVES

These valves are located between the carburetor and crankcase. The valve itself is a thin sheet of springy alloy steel. It springs open to allow fuel mixture to enter the crankcase, and springs closed to seal the crankcase.

Fig. 3-27 A Piston for a Two-cycle Engine Showing the Exhaust Side and the Intake Side.

There may be only one reed valve or there may be several reed valves working together. By far the greatest number of small gasoline engines operating on the two-cycle principle use reed valves even though it is possible to use a poppet valve in the crankcase.

OTHER TWO-CYCLE OPERATION

There are other methods of two-cycle operation; not all two-cycle engines use reed or leaf valves. These other engines would still have the two strokes, Compression and Power, with intake and exhaust taking place between the two, but different valves are used to achieve the result.

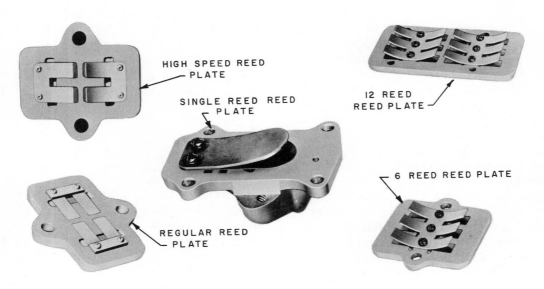

Fig. 3-28 Reed Plates.

ROTARY VALVE

Some two-cycle engines will be found that use a rotary valve for admitting fuel mixture. The rotary valve is a flat disc, with a section removed, that is fastened to the crankshaft. It normally seals the crankcase, but as the piston nears the top of its stroke and there is a slight vacuum in the crankcase, the valve is rotated to its open position, allowing fuel mixture to travel from the carburetor through the open valve and into the crankcase.

POPPET VALVES

Also, there are some two-cycle engines that use poppet valves to admit new fuel mixture into the crankcase. These poppet valves may be spring loaded and operated by differences in crankcase pressure. They open when the partial vacuum in the crankcase overcomes a slight spring tension. This happens when the piston is on its upward stroke. With the poppet valve open, fuel mixture rushes into the crank-

case. The poppet valve may also be operated by cam action. In this case a crankshaft cam opens the valve for the piston's upward stroke.

LOOP SCAVENGING

The loop scavenged two-cycle engine is basically the same in operation as other two-cycle engines. However, its piston is different in that it has large bores on either side of the piston skirt and the top is flat. The piston is forced down the cylinder on the power stroke. Exhaust ports are uncovered near the bottom of the stroke and exhaust starts out. Also, the large bores in the cylinder line up with intake passages and ports in the cylinder permitting new fuel mixture to rush in. The fuel mixture comes in from both sides, looping up to the top of the cylinder and then back down to scavenge out the remaining exhaust. Loop scavenging provides a more complete removal of exhaust gases and produces somewhat more horsepower per unit weight.

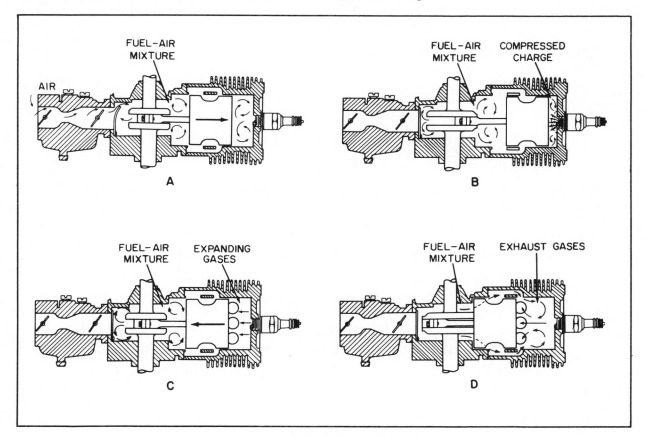

Fig. 3-29 Loop Scavenged Two-cycle Operation.

GENERAL STUDY QUESTIONS

1. Explain the term "Internal Combustion".

2. List the most essential parts of the engine.

3. Of what metal is the cylinder made?

4. What purpose does a gasket serve?

5. What is a combustion chamber?

6. What purpose do piston rings serve?

7. What is the function of the crankshaft?

8. What does the flywheel accomplish?

9. Explain how the valves are operated.

10. At what speed does the camshaft revolve relative to the crankshaft?

FOUR-STROKE CYCLE STUDY QUESTIONS

1. Explain what causes the fuel mixture to rush into the cylinder during the intake stroke.

2. What is accomplished on the compression stroke?

3. Explain the power stroke.

4. How many revolutions are required for the complete cycle?

5. Explain compression ratio.

TWO-STROKE CYCLE STUDY QUESTIONS

1. What parts are replaced by the intake and exhaust ports?

2. How is the two-stroke cycle piston different from the four-stroke cycle piston?

3. Explain the action of the reed or leaf valves.

4. Explain how intake and exhaust take place on a two-cycle engine.

5. What part does the crankcase play in regard to the fuel mixture?

6. How many revolutions are required for the complete cycle?

CLASS DISCUSSION TOPICS

● What other "systems" does an engine need in addition to the basic moving parts?

● Discuss the basic differences in two- and four-cycle engines.

● Discuss advantages and disadvantages of two- and four-cycle engines.

● Discuss how parts can be damaged by careless handling.

CLASS DEMONSTRATION TOPICS

► Inspect the basic parts of the two-cycle and four-cycle engine.

► Remove the cylinder head of a four-cycle engine, rotate the crankshaft, and observe the action of the valves.

► Remove the valve spring cover of a four-cycle engine, rotate the crankshaft, and observe the action of the valve lifters and valve springs.

► Remove the carburetor of a two-cycle engine, turn the engine over rapidly, and observe the action of the leaf valves. Note: this action is not pronounced and must be carefully observed.

► Using a disassembled engine, connect piston, connecting rod, and crankshaft to illustrate how reciprocating motion is converted to rotary motion.

► Using a camshaft and crankshaft, point out timing marks.

► Using a camshaft and crankshaft, count the gear teeth to illustrate the speed relationship of the two gears.

FUEL SYSTEMS, CARBURETION AND GOVERNORS

The fuel system must maintain a constant supply of gasoline for the engine, and the carburetor must correctly mix the gasoline and air together to form a combustible mixture which will burn rapidly when ignited in the combustion chamber.

A typical fuel system contains a <u>gasoline tank</u> (the reservoir for gasoline); a <u>carburetor</u> (a mixing device for gasoline); the <u>fuel line</u> (tubes made of rubber or copper through which the gasoline passes from the gas tank to the carburetor); and an <u>air cleaner</u> (a device for filtering air brought into the carburetor). In addition, the system may have a <u>shutoff valve</u> (a valve at the gas tank that can cut off the gasoline supply when the engine is not in use); a <u>fuel pump</u> (a pump that supplies the carburetor with a constant supply of gasoline); a <u>sediment bowl</u> (a small glass bowl attached to the fuel line where dirt and other foreign matter can settle out); and a <u>strainer</u> (a fine screen in the gas tank to prevent leaves and dirt from entering the fuel line).

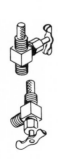 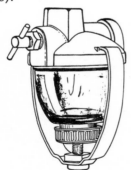

Fig. 4-1 Fuel Shutoff Valve

Fig. 4-2 Sediment Bowl with Built-in Fuel Shutoff Valve.

It should be realized that not all fuel systems will contain all eight of the basic parts: shutoff valves, sediment bowls, fuel pumps, and air cleaners are not common to all engines.

A constant supply of gasoline must be available at the carburetor. To provide this supply, four methods are in common use today: (1) Suction, (2) Gravity, (3) Fuel Pump, (4) Pressurized Tank.

SUCTION SYSTEM

The suction system is probably the simplest. With this method the gas tank is located below the carburetor and the gasoline is simply sucked up into the carburetor. However, the gas tank cannot be very far away from the carburetor or the carburetor action will not be strong enough to pull the gasoline from the tank.

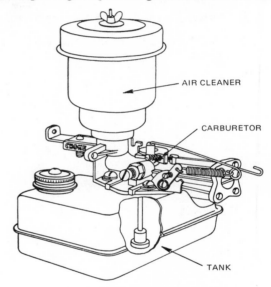

Fig. 4-3 Suction Feed Fuel System.

GRAVITY SYSTEM

With the gravity system the gas tank is located above the carburetor and the gasoline runs down hill to the carburetor. To prevent gasoline from continuously pouring through the carburetor, the carburetor has incorporated within it a float and float chamber. This float chamber provides a constant level of gasoline without flooding the carburetor. When gasoline is used, the float goes down, opening a valve to admit more gasoline to the float chamber; the float rises and shuts off the gasoline when it reaches its correct level. In actual practice the float and float valve do not rapidly open and close, but "assume a position" allowing the correct amount of fuel to constantly enter the float chamber. If the engine were speeded up, a new position would be assumed by the float and float valve, supplying an increased flow of gasoline.

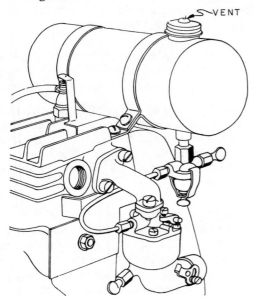

Fig. 4-4 Gravity Feed Fuel System.

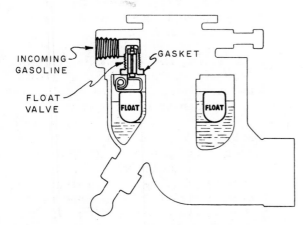

Fig. 4-5 Cutaway of a Carburetor, Showing Float and Float Valve.

FUEL PUMP

On many engines it is necessary to place the gas tank some distance from the carburetor, and, therefore, the fuel must be brought to the carburetor by some other means than gravity or suction. One method is by using a fuel pump. The automobile engine uses a fuel pump; among smaller engines, the outboard motor with a remote fuel tank is a good example of an engine that commonly uses a fuel pump.

The fuel pump used on many two-stroke cycle outboard motors is quite simple, consisting of a chamber, inlet and discharge valve, a rubber diaphragm, and a spring. This fuel pump is operated by the crankcase pressure. As the piston goes up, low pressure in the crankcase pulls the diaphragm toward the crankcase, sucking gasoline through the inlet valve into the fuel chamber. As the piston comes down, pressure in the crankcase pushes the diaphragm away from the crankcase. When this happens, the intake valve closes and the discharge valve opens, allowing the trapped fuel to be forced to the carburetor.

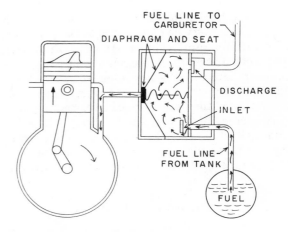

(A) Low Pressure in the Crankcase Allows the Fuel Chamber to be Filled.

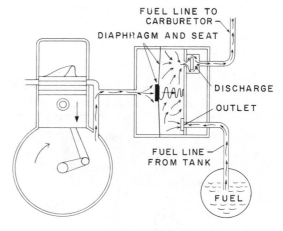

(B) High Pressure in the Crankcase Forces the Trapped Fuel on to the Carburetor.

Fig. 4-6 Effects of Low and High Pressure in Crankcase.

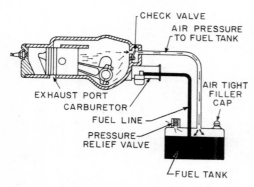

Fig. 4-7 Pressurized Fuel Systems can be Found on Many Outboard Engines.

PRESSURIZED FUEL SYSTEM

Another method for forcing fuel to travel long distances to the carburetor is the pressurized fuel system. This method is often used with outboard motors. Engines using this system will have two hoses or lines between the gas tank and the engine: one brings gasoline to the carburetor; the other brings air, under pressure, from the crankcase to the gas tank. If this method is used, the gas tank must be airtight so that sufficient air pressure can build up to force the gasoline to flow to the carburetor.

THE CARBURETOR

The carburetor must prepare a mixture of gasoline and air in the correct proportions for burning in the combustion chamber. The carburetor must function correctly under all

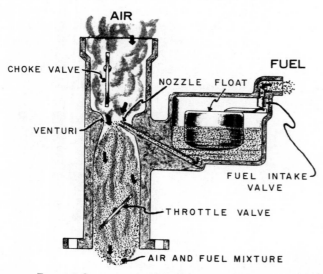

Fig. 4-8 Cross-section of a Simplified Carburetor.

engine speeds, under varying engine loads, in all weather conditions, and at all engine temperatures. To meet all these requirements, carburetors have many built-in parts and systems and may be quite complex.

Before studying the carburetor's operation in detail, an examination of the basic carburetor parts and their functions will be helpful.

Throttle or Butterfly: The throttle controls the speed of the engine by controlling the amount of fuel mixture that enters the combustion chamber. The more fuel mixture admitted, the faster the engine speed.

Choke: The choke controls the air flow into the carburetor. It is used only for starting the engine. In starting the engine, the operator closes the choke, cutting off most of the engine's air supply. This produces what is called a "rich mixture" (one containing a higher percentage of gasoline) which ignites and burns readily in a cold engine. As soon as the engine starts, the choke is opened.

Needle Valve: The needle valve controls the amount of gasoline that is available to the carburetor. It controls the richness or leanness of the fuel mixture.

Idle Valve: The idle valve also controls the amount of gasoline that is available to the carburetor but it functions only at low speeds or idling. Some carburetors have what is called a Slow-Speed Needle Valve which performs essentially the same job as the idle valve.

Float and Float Bowl: The float and float bowl will be found on all carburetors except the suction-fed carburetor and diaphragm carburetors (to be discussed later). The float and float bowl maintain a constant gasoline level in the carburetor.

Venturi: The venturi is a section of the carburetor that is constricted, or has a smaller cross-sectional area for the air to flow through. In the venturi, the gasoline and air are brought together and here their mixing begins.

<u>Jets</u>: Carburetor jets are small open-ings through which gasoline passes within the carburetor.

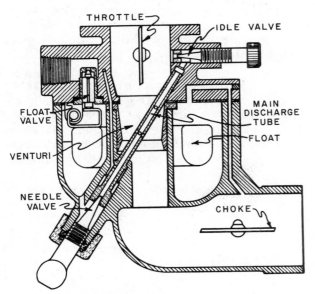

Fig. 4-9 A Gravity Fed Briggs-Stratton Carburetor.

In the beginning, air flows through the carburetor because there is a partial vacuum or suction in the combustion chamber as the piston travels down the cylinder. Normally we think of the air being sucked into the engine by the piston action. Atmospheric pressure out-side the engine actually pushes the air through the carburetor to equalize the lower pressure that is in the combustion chamber.

Air flows rapidly through the carburetor as the piston moves down, and in the carburetor the air must pass through a constriction called the "venturi". For the same amount of incom-ing air to pass through this smaller opening, it must travel faster, and this it does. Here a principle of physics comes into use: the greater the velocity of air passing through an opening, the lower the static air pressure exerted on the walls of the opening. The venturi creates a low-pressure area within the carburetor.

Also the principle of the air foil is used to gain lower pressure conditions in the venturi section. A fuel supply tube or jet is placed in the venturi section. The action of the incoming air causes a high pressure on the front of the jet but a very low pressure on the back of the jet. Gasoline is available in this jet and streams

out of the jet because a low-pressure area has been created by the action of the venturi and the air foil, and because the gasoline is under at-mospheric pressure which is greater. Greater pressure pushes the gasoline out of the dis-charge jet into the air stream.

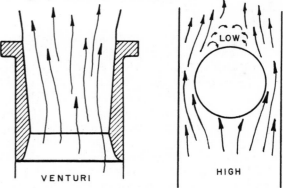

Fig. 4-10 (A) The Venturi Creates a Low Pressure in the Carburetor. (B) The Principle of the Air Foil Also Helps Create a Low Pressure.

As gasoline streams into the air flow, it is mixed thoroughly with the air. The best mixture of gasoline and air is 14 or 15 parts of air to 1 part of gasoline, by weight. This air to gasoline ratio can be changed for differ-ent operating conditions; heavy load and fast acceleration require more gasoline (richer mixture). The needle valve is used to change this ratio; it controls amount of gasoline that is available to be drawn from the discharge jet.

The throttle or butterfly is mounted on a shaft beyond the venturi section. The operator of the engine controls its setting to control the engine speed. When the throttle is wide open the butterfly does not restrict the flow of air; air flows easily through the carburetor; the engine is operating at its top speed. As the op-erator closes the throttle, the flow of air is restricted. A smaller amount of air can rush through the carburetor, therefore, the air pres-sures at the venturi section are not as low and less gasoline streams from the discharge holes. With less fuel mixture in the combustion cham-ber, the piston is pushed down with less force during combustion; power and speed are reduced.

The ratio of air to fuel remains approxi-mately the same through the different throttle settings. However, when the throttle is closed and the engine begins to idle, very little air is drawn through the carburetor and the difference between atmospheric pressure and venturi air is slight. Little gasoline is drawn from the discharge jet. In fact, the mixture of gasoline is so "lean" that a special idling device must be built into the carburetor to provide a "richer" mixture for idling.

In some carburetors, the main discharge jet is continued on up past the venturi section to the area of the throttle. It discharges fuel into a small well and jet that are behind the throttle butterfly when it is closed. The air pressure behind the butterfly is very low. Therefore, gasoline streams from the idle jet readily and mixes with the small amount of air that is coming through the carburetor and a "rich" fuel mixture is provided for idling. A threaded needle valve called the "idle valve" controls the amount of gasoline that can be drawn from the idle jet.

When a cold engine is to be started, an extremely rich mixture of gasoline and air must be provided if the engine is to start easily. The choke will provide this rich mixture. It is a butterfly placed in the air horn before the venturi section. For starting, the choke is closed, shutting off most of the carburetor's air supply. When the engine is turned over slowly, usually by hand, little air is drawn through the carburetor but the air pressure within the carburetor is very low and gasoline streams from the main discharge jet, mixing with the air that does get by the choke. As soon as the engine starts, the choke is opened. The fuel mixture provided when the engine is choked is actually so rich that all the gasoline may not vaporize with the air and "raw" liquid gasoline may be drawn into the combustion chamber. Continued operation with the choke closed may cause crankcase dilution, i.e., the raw gasoline seeps into the crankcase diluting the lubricating oil.

This is now the complete carburetor: (1) Float and float bowl, maintaining a constant reservoir of gasoline in the carburetor; (2) Venturi section, producing a low-pressure area; (3) Needle valve, controlling the richness or leanness of fuel mixture; (4) Main discharge jet, squirting gasoline into airstream; (5) Idle valve, providing a rich mixture for idling conditions; (6) Choke, producing an extremely rich mixture for easy starting.

DIAPHRAGM CARBURETOR

This type of carburetor has come into wide use, especially on chain saws. It is also found on other applications where the engine may be tipped at extreme angles. The diaphragm supplies the carburetor with a constant supply of gasoline.

The diaphragm carburetor may be gravity fed. Crankcase pressure moves the diaphragm and associated linkages to allow gasoline to enter the fuel chamber. The gasoline is available by gravity and the in and out motion of the diaphragm meters out fuel in the correct amount.

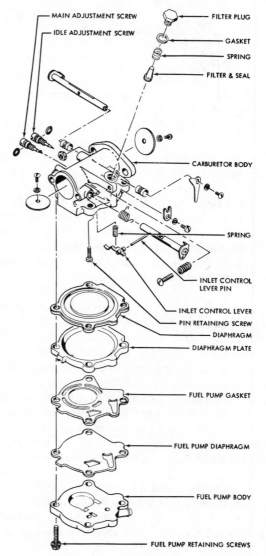

Fig. 4-11 An Exploded View of a Diaphragm Carburetor with Built-in Fuel Pump.

Diaphragm carburetors are also built which operate with a built-in fuel pump. The fuel pump operates on crankcase pressure and its action is virtually the same as that of the outboard fuel pump previously discussed. Diaphragm carburetors have found wide acceptance in the small engine field.

1   HIGH SPEED MIXTURE ADJUSTING SCREW
2   IDLE SPEED MIXTURE ADJUSTING SCREW
3   IDLE SPEED REGULATING SCREW

Fig. 4-12 Diaphragm Carburetor Installed on a McCulloch Chain Saw.

HIGH-SPEED — LOW-SPEED CONTROLS

Many carburetors, especially those on outboard motors, have what is termed "high-speed" and "low-speed" controls. The high-speed control corresponds to the main needle valve while the slow-speed control corresponds to the idle valve. The carburetors used on Johnson motors, as well as those used on other manufacturers' motors, have these controls. When the throttle butterfly is closed or just about closed, the motor is idling or operating at slow speeds. Now the fuel comes from the slow-speed jet because it is located in the area of greatest suction. As the butterfly opens, the low-pressure or suction effect is felt more and more on the high-speed jet located in the venturi until the low-speed jet becomes ineffective. Now the great bulk of the fuel is coming from the high-speed jet.

It would not be correct to say that the two controls are independent of each other. Rather, there is a shifting of control throughout the speed range. One does not abruptly "cut-out" and the other "cut-in". Even at full throttle, a small amount of gasoline can be entering through the slow-speed jet.

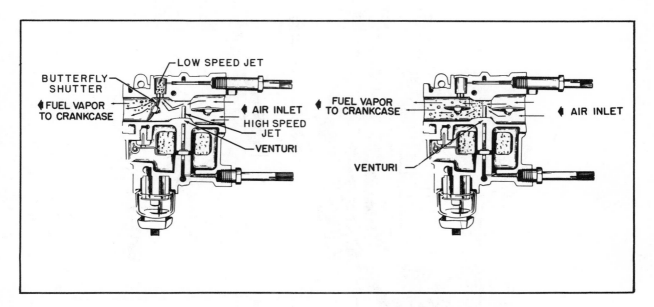

(A) Showing Butterfly Shutter Set for Slow Speed Operation (Closed). Note Maximum Fuel Vaporization at Slow Speed.

(B) Butterfly Shutter Full Open for High Speed Performance. Note Maximum Fuel Vaporization at High Speed Jet with Minimum of Vaporization at the Slow Speed Jets.

Fig. 4-13 Sectional View of a Mixing Chamber

AIR BLEEDING CARBURETORS

There is a tendency for carburetors to supply too rich a fuel mixture at high speeds; the ratio of gasoline to air increases as the velocity of the air passing through the carburetor increases. One method that is commonly used to correct this condition is Air Bleeding. The Zenith carburetor shown below uses the principle of air bleeding.

A small amount of air is introduced into the main discharge well vent to restrict the flow of gasoline from the main discharge jet. As engine speeds are increased, greater amounts of air are brought into the main discharge well vent, placing a greater restriction on the gasoline flow, thereby overcoming the carburetor's natural tendency to provide too rich a mixture at high speeds. This action maintains the proper ratio of fuel and air between a throttle setting of one-fourth to wide open.

The air that enters the discharge well vent does mix with the gasoline and is drawn through the main discharge jet into the main air stream.

ACCELERATING PUMP

Another problem inherent in all carburetors is a response lag when the throttle is quickly opened. Air can react very quickly to an increased demand but gasoline lags behind. The result is too lean a mixture and acceleration is sluggish. Carburetors equipped with an accelerating pump provide instant response for rapid acceleration.

The main parts of the accelerating pump are the spring, vacuum piston, and fuel cylinder. At idling and low operating speeds, the vacuum piston is drawn to the top of the fuel cylinder by the engine vacuum, which is strong enough to overcome the force of the spring, holding the piston at the top of its stroke. Now the fuel cylinder is filled with gasoline.

If the throttle is suddenly opened, the engine vacuum drops enough for the piston spring to overcome the force of the vacuum, pushing the pump piston down the fuel cylinder. This reserve amount of gasoline in the fuel cylinder is forced into the main discharge jet and on into the carburetor venturi.

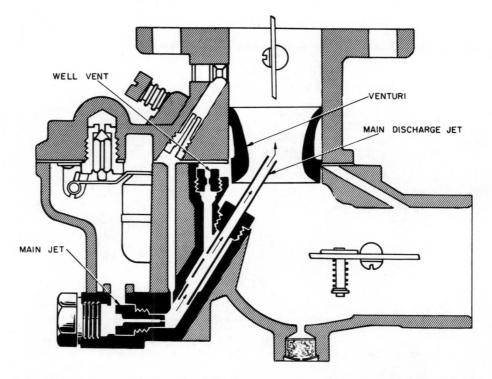

Fig. 4-14 Air Brought into the Well Vent Bleeds into the Main Discharge Jet, Maintaining Correct Air-Fuel Ratio throughout Throttle Range.

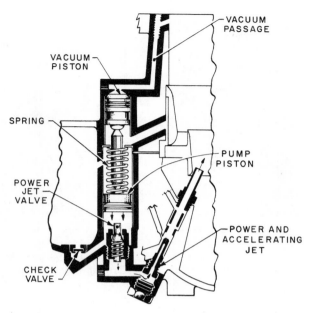

Fig. 4-15 Cross-section of an Accelerating Pump.

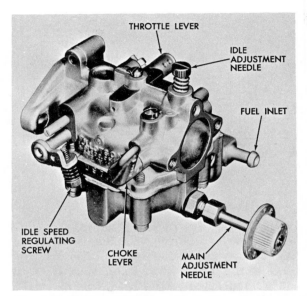

Fig. 4-16 Tillotson MD Float Feed Type Carburetor.

CARBURETOR ADJUSTMENTS

The engine manufacturer will set the carburetor adjustments at the factory. These settings will cover normal operation. However, after a long period of usage, or under special operating conditions, it may be necessary to adjust the carburetor. In adjusting the carburetor, the main needle valve and the idle valve are both reset to give the desired richness or leanness of fuel mixture. Too lean a mixture can be detected by the engine missing and backfiring, while too rich a mixture can be detected by heavy exhaust and sluggish operation.

Here is a typical procedure for adjusting a carburetor for maximum power and efficiency.

1. Close the main needle valve and idle valve "finger tight". Excessive force can damage the needle valve. Turn clockwise to close.

2. Open the main needle valve one turn. Open the idle valve 3/4 turn. Turn counterclockwise to open.

3. Start the engine, open the choke, and allow the engine to reach operating temperature.

4. Run engine at operating speed (2/3 to 3/4 of full throttle). Turn the main needle valve in (clockwise) until the engine slows down, indicating too lean a mixture. Note the position of the valve. Turn the needle valve out (counterclockwise) until the engine speeds up and then slows down, indicating too rich a mixture. Note the position of the valve. Reposition the valve halfway between the rich and lean settings.

5. Close the throttle so the engine runs slightly faster than normal idle speed. Turn the idle valve in (clockwise) until the engine slows down, then turn the idle valve out until the engine speeds up and idles smoothly. Adjust the idle-speed regulating screw to the desired idle speed.

NOTE: Idle speed is not the slowest speed at which the engine will operate; rather, it is a slow speed that maintains good air flow for cooling and a good takeoff spot for even acceleration. A tachometer and the manufacturer's specifications regarding proper idle speed are necessary for the best adjustment.

6. Test the acceleration of the engine by opening the throttle rapidly. If acceleration is sluggish, a slightly richer fuel mixture is usually needed.

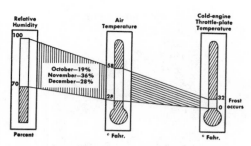

Fig. 4-17 Humidity and Temperature Conditions Which May Lead to Carburetor Icing.

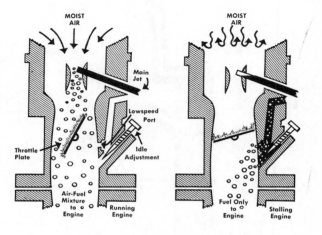

Fig. 4-18 Under Carburetor Icing Conditions, Ice Forms on the Throttle Plate, Cutting off Air to the Engine When Throttle Closes.

CARBURETOR ICING

Carburetor icing is an annoying phenomenon that may occur when the engine is cold and certain atmospheric conditions are present. If the temperature is between 28°F. and 58°F., and the relative humidity is above 70%, carburetor icing may take place.

Fuel mixture of gasoline and air rushes through the carburetor and the rapid action of evaporating gasoline chills the throttle plate to about 0°F. Moisture in the air will condense and freeze on the throttle plate when the relative humidity is high. This formation of ice restricts the air flow through the carburetor

and at low or idle settings the ice may completely block off the air flow, stalling the engine.

When this condition is present, the engine can be restarted but will stall again at low or idle speeds. As soon as the carburetor is warm enough to prevent ice formation, normal operation can take place.

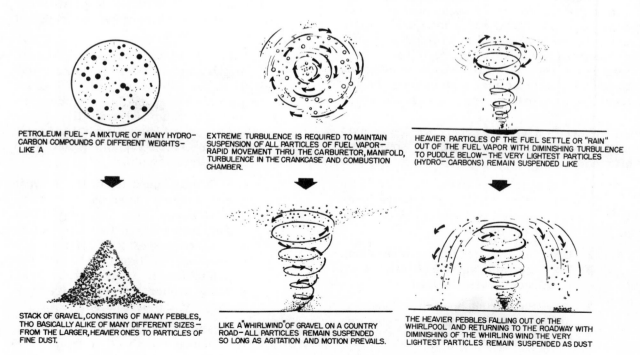

PETROLEUM FUEL – A MIXTURE OF MANY HYDRO-CARBON COMPOUNDS OF DIFFERENT WEIGHTS – LIKE A

EXTREME TURBULENCE IS REQUIRED TO MAINTAIN SUSPENSION OF ALL PARTICLES OF FUEL VAPOR – RAPID MOVEMENT THRU THE CARBURETOR, MANIFOLD, TURBULENCE IN THE CRANKCASE AND COMBUSTION CHAMBER.

HEAVIER PARTICLES OF THE FUEL SETTLE OR "RAIN" OUT OF THE FUEL VAPOR WITH DIMINISHING TURBULENCE TO PUDDLE BELOW – THE VERY LIGHTEST PARTICLES (HYDRO-CARBONS) REMAIN SUSPENDED LIKE

STACK OF GRAVEL, CONSISTING OF MANY PEBBLES, THO BASICALLY ALIKE OF MANY DIFFERENT SIZES – FROM THE LARGER, HEAVIER ONES TO PARTICLES OF FINE DUST.

LIKE A "WHIRLWIND" OF GRAVEL ON A COUNTRY ROAD – ALL PARTICLES REMAIN SUSPENDED SO LONG AS AGITATION AND MOTION PREVAILS.

THE HEAVIER PEBBLES FALLING OUT OF THE WHIRLPOOL AND RETURNING TO THE ROADWAY WITH DIMINISHING OF THE WHIRLING WIND THE VERY LIGHTEST PARTICLES REMAIN SUSPENDED AS DUST

Fig. 4-19 Turbulence is Necessary to Keep the Gasoline Molecules Suspended in the Air.

VAPOR LOCK

Vapor lock can occur anywhere along the fuel line, fuel pump, or in the carburetor when temperatures are high enough to vaporize the gasoline. Gasoline vapor in these places will cut off the liquid fuel supply, stalling the engine. If vapor lock occurs, the operator must wait until the carburetor, gas line, and fuel pump cool off and the gasoline vapor returns to liquid before the engine will restart. Vapor lock usually occurs on unseasonably hot days and is more troublesome at high altitudes.

GASOLINE

Most internal combustion engines burn gasoline as their fuel. Gasoline comes from petroleum, also called crude oil. Crude oil is actually a mixture of different hydrocarbons; gasoline, kerosene, heating oil, lubricating oil, and asphalt. These chemicals are all hydrocarbons, but they have characteristics that are quite different. Hydrogen is a light, colorless, odorless gas; carbon is black and solid. Different combinations of carbon and hydrogen give the different characteristics of hydrocarbon products.

The various hydrocarbons are separated by distillation of the crude oil. Crude oil is first heated to a temperature of 700° to 800° F. and then released into a fractionating or bubble tower, Fig. 4-20. The tower contains 20 to 30 trays through which hydrocarbon vapors can rise from below. When the heated crude oil is released into the tower, most of it flashes into vapor. As the vapors rise, they cool, and each type of hydrocarbon condenses at a different tray level. Heavier hydrocarbons condense first át relatively high temperatures. The lighter hydrocarbons, such as gasoline, condense high in the tower at relatively low temperatures. Gasoline is then further processed to improve its qualities.

Good gasoline must have several characteristics. It must vaporize at low temperatures for good starting. It must be low in gum and sulfur content. It must not deteriorate during storage. It must not knock in the engine. It must have proper vaporizing characteristics for the climate and altitude. It must burn cleanly to reduce air polution.

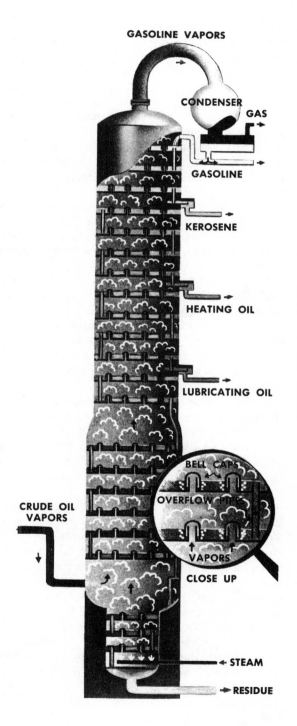

Fig. 4-20 Distillation of Petroleum in the "Bubble Tower."

The vaporizing ability of gasoline is the key to its success as a fuel. In order to burn inside the engine, the hydrocarbon molecules of the gasoline must be mixed with air since oxygen is also necessary for combustion. To mix the gasoline and air, there must be rapid motion and turbulence to keep the molecules suspended in the air. Of course, there is rapid motion and turbulence in the carburetor, and this turbulence continues on into the combustion chamber. In fact, combustion chambers are designed to create the maximum turbulence so that gasoline molecules will stay suspended in the air until they are ignited. A thorough mixture insures smooth and complete burning of gasoline, delivering maximum power.

Although the vaporizing ability of gasoline makes it an excellent fuel, it also presents a considerable fire and explosion hazard. When vaporized, a gallon of gasoline produces 21 cubic feet of vapor. If this vapor were combined with air in a mixture of 1.4 to 7.6 percent gasoline by volume, it would explode when ignited. Therefore, only one gallon of gasoline properly vaporized would completely fill an average living room with explosive vapor. The safety rules given in unit 8 for the use and storage of gasoline should always be followed.

In engines the compression of the fuel mixture of gasoline and air results in high combustion pressures and the force necessary to move the piston. This compression may also cause the gasoline to explode (detonate) in the engine instead of burning smoothly.

Detonation is also called knocking, fuel knock, spark knock, carbon knock, and ping. To understand it, one has to think in slow motion. The spark plug ignites the fuel mixture, and a flame front moves out from this starting point. As the flame front sweeps across the combustion chamber, heat and pressure build. The as-yet-unburned portion of the fuel mixture ahead of the flame front is exposed to this heat and pressure. If it self-detonates, two flame fronts are created which race toward each other. The last unburned portion of fuel caught between the two fronts explodes with hammerlike force. Detonation causes a knocking sound in the engine and power loss. Repeated detonation can damage the piston, Fig. 4-21.

Preignition causes the same undesirable effects as detonation, including engine damage, Fig. 4-22. The cause of preignition, however, is somewhat different. Hot spots, or red carbon deposits, actually ignite the fuel mixture and begin combustion before the spark plug fires.

Fig. 4-21 Detonation Damage.

Fig. 4-22 Preignition Damage.

The antiknock quality of a gasoline, or its ability to burn without knocking, is called its octane rating. Gasoline that has no knocking characteristics at all is rated at 100. Usually, tetraethyl lead is added to gasoline to give it the proper antiknock quality. If lead is not used, as in nonleaded gasoline, then special aromatic compounds are used to obtain smooth burning.

Spark occurs.....

Spark occurs.....

Ignited by hot deposit..

...continues rapidly...

......continues......

..regular ignition spark..

..combustion begins..

..combustion begins..

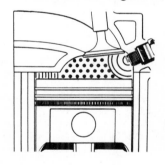

..ignites remaining fuel.

...and is completed.

......detonation.

..flame fronts collide.

Fig. 4-23A Normal Combustion

Fig. 4-23B Detonation

Fig. 4-23C Preignition

The higher the compression ratio of the engine, the higher the octane requirements for the engine's fuel. Low-grade gasoline with an octane rating of 70-85 is suitable for compression ratios of 5-7 to 1. Regular grade gasoline with an octane rating of 88-94 is suitable for compression ratios of 7-8.5 to 1. Most small engine manufacturers recommend the use of regular gasoline. High-octane, premium gasoline does not improve the performance of small engines. Premium gasoline with an octane rating of about 100 is suitable for compression ratios of 9-10 to 1. Super premium has an octane rating of over 100 and is good for engines with compression ratios of 9.5-10.5 to 1.

ENGINE SPEED GOVERNORS

Speed governors are used to keep engine speed at a constant rate regardless of the load. For instance, a lawnmower is required to cut tall as well as short grass. A governor will insure that the engine operates at the same speed in spite of the varying load conditions. Engines on many other applications need this "constant speed" feature too.

Speed governors are also used to keep the engine operating speed below a given, preset rate, so that the engine speed will not surpass this rate. This maximum rate is established and the governor set accordingly by the engineers at the factory. This type of speed governor protects both the engine and the operator from speeds that are dangerously high.

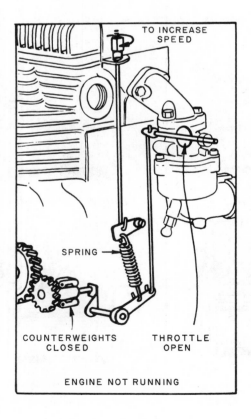 

Fig. 4-24 Briggs-Stratton Mechanical Governor Operates on Centrifugal Force.

There are two main governor systems used on small gasoline engines: (1) Mechanical or Centrifugal type and (2) Pneumatic or Air Vane type. Although there are other types of governors, most of the governors used on small engines will fall into one of these categories.

MECHANICAL GOVERNORS

The mechanical governor operates on centrifugal force. With this method, counterweights mounted on a geared shaft, a governor spring, and the associated governor linkages, maintain the engine speed at the desired r.p.m's. For constant speed operation the action follows this pattern: when the engine is stopped, the mechanical governor's spring will pull the throttle to an open position. The governor spring tends to keep the throttle open, but as engine speed increases, centrifugal force throws the "hinged" counterweights further and further from their shaft. This action puts tension on the spring in the other direction, to close the throttle. At governed speed the spring tension is overcome by the counterweights and the throttle will open no further, a balanced position is maintained; the engine assumes its maximum speed.

If, however, a greater load is placed on the engine, its speed will slow down; the hinged counterweights will swing inward due to lessened centrifugal force; and the governor spring will become dominant, opening the throttle wider. With a wider throttle opening, the engine speeds up until governed speed is reached again. The action described is fast and smooth; little time is needed for the governor to meet "revised" load conditions.

The governed speed can be changed somewhat by varying the tension on the governor spring. The more tension there is on the governor spring, the higher the governed speed.

Another commonly used centrifugal or mechanical governor is the flyball type. With this governor, round steel balls in a spring loaded raceway move outward as the engine speed increases. The centrifugal force applied by these balls increases until it balances the spring tension holding the throttle open. At this point, the governed speed is reached.

Fig. 4-25 A Centrifugal Governor (Mechanical) Used on Certain Wisconsin Engines.

Fig. 4-26 A Mechanical Governor Used on Certain Lauson Engines.

PNEUMATIC OR AIR VANE GOVERNORS

Many engines will be seen with a pneumatic or air vane governor. Again, the governor works to hold the throttle at the governed speed position, preventing it from opening further. An air vane, which is located near the flywheel blower, controls the speed. As engine speed increases, the flywheel blower pushes more air against the air vane, causing it to change position. At governed speed the air vane position overcomes the governor spring tension and a balance is assumed.

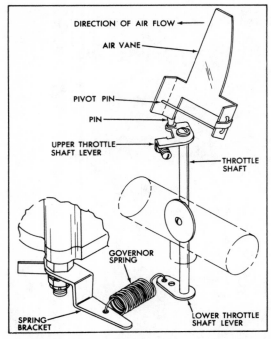

Fig. 4-27 A Pneumatic or Air Vane Governor Used on Certain Lauson Engines.

Engines are often designed so that they may be accelerated freely from idle speed on up to governed speed. If this is the case, the governor spring takes on very little tension at low speeds. The operator has full control. As the throttle is opened further and further, there is more tension on the spring, both from the operator and from the strengthening governor action. At governed speed, the force to close the throttle balances the spring tension to open the throttle.

It should be noted that not all engines are equipped with engine speed governors. Engines that will always be operated under a near constant load will find the governor unnecessary. The load of an outboard engine is a good example of a near constant load. Although different loads are placed on the outboard (number of people in boat, position of people in boat, use of trolling mechanism to decrease boat's speed, etc.), the load is constant at full throttle for a given period of use.

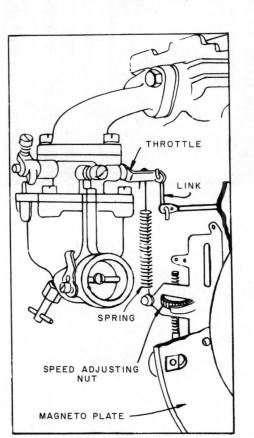

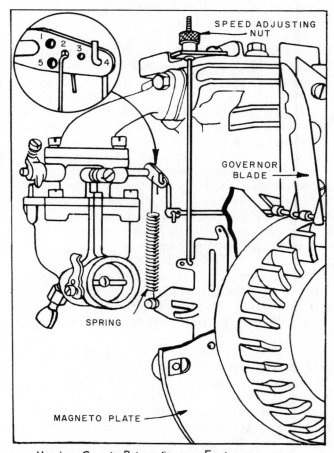

Fig. 4-28 A Pneumatic or Air Vane Governor Used on Certain Briggs-Stratton Engines.

GENERAL STUDY QUESTIONS

1. Name the basic parts of a typical fuel system.

2. What force is used to operate most outboard fuel pumps?

3. How can a pressurized fuel system be recognized?

4. Briefly describe the task of the carburetor.

5. Name the main parts that make up the carburetor.

6. What causes air to flow through the carburetor?

7. What causes low air pressure in the venturi section of the carburetor?

8. What is the function of the throttle?

9. What is the function of the choke?

10. What is the function of the needle valve?

11. Explain the action of the float in a float-type carburetor.

12. What is the advantage of a diaphragm carburetor?

13. Explain the principle of air bleeding.

14. Explain the operation of the accelerating pump.

15. What is a "rich mixture"?

16. Why is turbulence important inside the combustion chamber?

17. What two main purposes do engine speed governors serve?

18. What are the two main classifications of engine speed governors?

19. Are engines made that do not have engine speed governors?

20. Can the governed speed be changed? How?

CLASS DEMONSTRATION TOPICS

◆ Using a fully assembled engine, have the students trace the flow of fuel and also identify the parts of the fuel system.

◆ Demonstrate, with an engine operating, how to set a carburetor for maximum power.

◆ Demonstrate, with an operating engine, how "too lean" or "too rich" mixture affects operation and acceleration.

◆ Demonstrate the action of a carburetor with an atomizer or insect sprayer.

◆ Disassemble and inspect a fuel pump (automotive or outboard).

◆ Illustrate vaporization with a small saucer of gasoline beside a small saucer of oil.

◆ Demonstrate, with an operating engine, how the governor maintains governed speed and will not allow the throttle to open to "full".

◆ Demonstrate how governed speed can be changed. Stop engine to readjust governor.

Unit 5

LUBRICATION

Whenever surfaces move against one another, they cause friction and friction results in heat and wear. Lubricating oils have one main job to perform in the engine; namely, to reduce friction. The lubricating oil provides a film that separates the moving metal surfaces and keeps the contact of metal against metal to an absolute minimum.

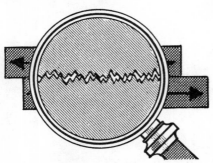

Fig. 5-1 Exaggerated View of Metal Surfaces in Contact.

Without a lubricating oil or with insufficient lubrication, the heat of friction builds up rapidly. Engine parts become so hot that they fail; that is, the metal begins to melt — it gets mushy; bearing surfaces seize, parts warp out of shape, or parts actually break. The common expression is to say the engine "burns up".

As an example of friction, take a book, lay it on the table and then push the book with your hand. Notice the resistance. Friction makes the book difficult to slide. Now place three round pencils between the book and the table top, push the book and notice how easily it moves. Friction has been greatly reduced. Oil molecules correspond to the pencils by forming a coating between two moving surfaces. With oil, the metal surfaces literally roll along on the oil molecules and friction is greatly reduced.

Besides reducing friction and the wear and heat it causes, the lubricating oil serves several other important functions.

1. Oil Seals Power

The oil film seals power, particularly between the piston and cylinder walls. The tremendous pressures in the combustion chamber cannot pass by the airtight seal the oil film provides. If this oil film fails, a condition called "blow-by" exists. Combustion gases push by the film and enter the crankcase. Blow-by not only reduces engine power, but also has a harmful effect on the oil's lubricating quality.

2. Oil Helps to Dissipate Heat

Oil helps to dissipate heat by providing a good path for heat transfer. Heat conducts readily from inside metal parts through an oil film to outside metal parts that are cooled by air or water. Also, heat is carried away as "new" oil arrives from the crankcase and the "hot" oil is washed back to the crankcase.

3. Oil Keeps the Engine Clean

Oil keeps the engine clean by washing away microscopic pieces of metal that have been worn off moving parts. These minute pieces settle out in the crankcase or they are trapped in the oil filter, if one is used.

4. Oil Cushions Bearing Loads

The oil film has a cushioning effect since it is squeezed from between the bearing surfaces relatively slowly. It has a shock absorber action. For example, when the power stroke starts, the hard shock of combustion is transferred to the bearing surfaces; the oil film helps to cushion this shock.

5. Oil Protects Against Rusting

The oil film protects steel parts from rusting. Air, moisture, and corrosive substances cannot reach the metal to oxidize or corrode the surface.

FRICTION BEARINGS

There are three types of <u>friction</u> <u>bearings</u> used in a small gasoline engine: journal, guide, and thrust. The <u>journal</u> bearing is the most familiar. This bearing supports a revolving or oscillating shaft. The connecting rod around the crankshaft and the main bearings are examples. The <u>guide</u> bearing reduces the friction of surfaces sliding longitudinally against each other such as the piston in the cylinder. The <u>thrust</u> bearing supports or limits the longitudinal motion of a rotating shaft.

Bearing inserts are commonly used on larger engines or heavy-duty small engines, particularly on the connecting rod and cap and the main bearings. The inserts are precision-made of layers of various metals and alloys. Alloys such as babbit, copper-lead, bronze, aluminum, cadmium, and silver are commonly used. Crankshaft surfaces are, of course steel. Most small engines, however, do not have bearing inserts, using just the aluminum connecting rod around the steel journal of the crankshaft.

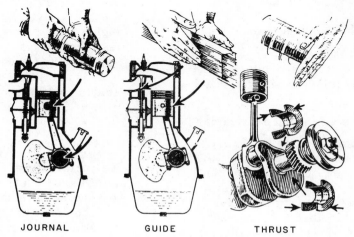

JOURNAL   GUIDE   THRUST

Fig. 5-2 Three Major Classifications of Friction Type Bearings.

ANTI-FRICTION BEARINGS

Anti-friction bearings are also commonly used in engines. They substitute rolling friction for sliding friction. Ball bearings, roller bearings and needle bearings are of this type. On many small engines the main bearings are of the anti-friction type.

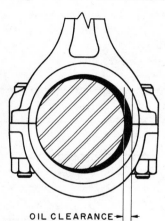

OIL CLEARANCE

Fig. 5-3 Oil Clearance between Journal and Bearing (Exaggerated).

TAPERED ROLLER BEARINGS

NEEDLE BEARINGS   BALL BEARINGS

Fig. 5-4 Types of Anti-Friction Bearings.

QUALITY DESIGNATION OF OIL AND
SAE NUMBER

Selecting engine lubricating oils can be confusing: one must be aware of different viscosities, different qualities, different refining companies, different additives, and the meaning of a multitude of advertising phrases. It's a lot of information to sort out and to understand. Generally, lubricating oils used for various small gasoline engines are the same ones that are used for automobile engines. However, two-stroke cycle engines almost always use special two-cycle oil.

Quality Designation by the American
Petroleum Institute

The API classifications are usually found on the top of the oil can and refer to its quality.

● Oils for Service SE (Service Extreme)

SE oils are suitable for the most severe type of operation beginning with 1972 models and some 1971 automobiles. Extreme conditions of start-stop driving; short trip, cold weather driving; and high speed, long distance, hot weather driving can be handled by this oil. The oil meets the requirements of automobiles that are equipped with emission control devices and engines operating under manufacturers' warranties.

● Oils for SD (Service Deluxe)

Most small engine manufacturers approve the use of SD (formerly MS) oil in their engines. These oils provide protection against high and low temperature engine deposits, rust, corrosion, and wear.

● Oils for Service SC

This oil was also formerly classified as being suitable for MS. It has much the same characteristics as SD but is not quite as effective as SD oil. SC oil was developed for auto engines of 1964-67, while SD oil was developed for engines of 1968-70 manufacture.

● Oils for Service SB (Formerly MM)

This oil is recommended for moderate operating conditions such as moderate speeds in warm weather; short distance, high speed driving; and alternate long and short trips in cool weather. It is satisfactory for certain older autos but not for new autos under warranty. It is seldom recommended for small engine use.

● Oils for Service SA (Formerly ML)

No performance requirements are set for this oil. It is straight mineral oil and may be suitable for some light service requirements. SA oil is not recommended by small engine manufacturers.

The new classifications for oils suitable for diesel service are CA, CB, CC, and CD; these replace the old classifications of DG, AM (Supp-1), DM (MIL-L2104B), and DS respectively.

It is wise to buy the best quality oil for your engine. Skimping on quality for a few cents savings may be more expensive in the long run due to engine wear.

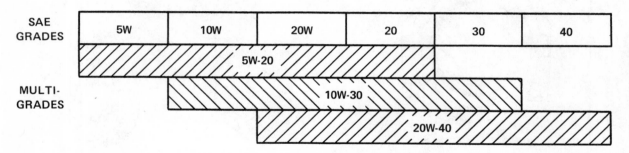

Fig. 5-5 Multigrade Motor Oils Span Several SAE (single) Grades.

Viscosity Classification by the Society of Automotive Engineers

The SAE (Society of Automotive Engineers) number of an oil indicates its viscosity, or thickness. Oils may be very thin (light), or they may be quite thick (heavy). The range is from SAE 5W to SAE 50; the higher the number, the thicker the oil. The owner should consult the engine manufacturer's instruction book for the correct oil. SAE 10W, SAE 20W, SAE 20, and SAE 30 are the most commonly used oil weights. Usually, manufacturers recommend SAE 30 for summer use and either SAE 10W or SAE 20W for cold, subfreezing weather. Winter use demands a thinner oil since oil thickens in cold weather. The "W", as in 5W, 10W, and 20W, indicates that the oil is designed for service in subfreezing weather.

● Multigrade Oils

Multigrade oils may also be used in small gasoline engines. These oils span several SAE classifications because they have a very high viscosity index. Typical classifications are SAE 5W-20 or 5W-10W-20; 5W-30 or 5W-10W-20W-30; SAE 10W-30 or 10W-20W-30; 10W-40, and 20W-40. The first number represents the low temperature viscosity of the oil; the last number represents its high temperature viscosity. For example, 10W-30 passes the viscosity test of SAE 10W at low temperatures and the viscosity test of SAE 30 at high temperatures. These oils are also referred to as all-season, all-weather oils, multiviscosity, or multiviscosity grade oils.

Motor Oil Additives

The best oil could not do its job properly in a modern engine if additives were not blended into the base oil.

● Pour-Point Depressants

Pour-Point depressants keep the oil liquid even at very low temperatures when the wax in oil would otherwise congeal into a buttery consistency to render the oil ineffective.

● Oxidation and Bearing Corrosion Inhibitors

These additives prevent the rapid oxidation of the oil by excessive heat. Without these inhibitors viscous, gummy materials are formed. Some of these oxidation products attack metals such as lead, cadmium, and silver which are often used in bearings. The inhibiting compounds are gradually used up and, thus, regular oil changes are needed.

● Rust and Corrosion Inhibitors

These inhibitors protect against the damage that might be caused by acids and water which are by-products of combustion. Basically, acids are neutralized by alkaline materials, much the same as vinegar can be neutralized with baking soda. Special chemicals surround or capture molecules of water, preventing their contact with the metal, while other chemicals with an extreme affinity for metal form an unbroken film on the metal parts. These inhibitors are also used up in time.

● Detergent/Dispersant Additives

This type of additive prevents the formation of sludge and varnish. Detergents work much the same as household detergents in that they have the ability to disperse and suspend combustion contaminants in the oil but do not affect the lubricating quality of the oil. When the oil is changed, all the contaminants are discarded with the oil so that the engine is kept clean. Larger particles of foreign matter in the oil either settle out in the engine base or are trapped in the oil filter if one is used.

● Foam Inhibitors

Foam inhibitors are present in all high quality motor oils to prevent the oil from being whipped into a froth or foam. The action in the crankcase tends to bring air into the oil, and foaming oil is not an effective lubricant. Foam inhibitors called silicones have the ability to break down the tiny air bubbles and cause the foam to collapse.

LUBRICATION OF FOUR-CYCLE ENGINES

Since all moving parts of the engine must be lubricated to avoid engine failure, a constant supply of oil must be provided. Each engine, therefore, carries its own reservoir of oil in its crankcase where the main engine parts are located.

Basically the oil is either pumped to or splashed on the parts and bearing surfaces that need lubrication. There are several lubrication systems used on small engines and the systems discussed below are among the most common.

1. Simple Splash
2. Constant Level Splash
3. Ejection Pump
4. Barrel-Type Pump
5. Full Pressure Lubrication

The Splash System is perhaps the simplest system for lubrication. It consists of a splasher or dipper that is fastened to the connecting rod cap. Each time the piston nears the bottom on its stroke the dipper splashes into the oil reservoir in the crankcase, splashing oil onto all parts inside the crankcase. Since the engine is operating at 2000 to 3000 r.p.m.'s, the parts are literally drenched by millions of oil droplets. Some engines use an oil slinger that is driven by the camshaft. The slinger performs a similar function to the dipper.

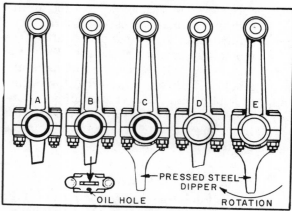

Fig. 5-6 Dippers used on Different Connecting Rods.

The Constant Level Splash System has three refinements over the simple splash system: (1) a pump, (2) a splash trough, and (3) a strainer. With this system a cam-operated pump brings oil from the bottom of the crank-

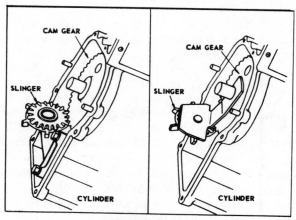

Fig. 5-7 Oil Slinger.

case into a splash trough. Again the splasher on the connecting rod dips into the oil, splashing it on all parts inside the crankcase. The strainer prevents any large pieces of foreign matter from recirculating through the system. The pump maintains a constant oil supply in the trough regardless of the oil level in the crankcase.

Fig. 5-8 Constant Level Splash System used on Some Models of Wisconsin Engines.

Ejection Pumps of various types are found on many small engines. With this method, a cam-operated pump draws oil from the bottom of the crankcase and sprays or squirts it onto the connecting rod. Some of the oil enters the connecting rod bearing through small holes, while the remainder is deflected onto the other parts within the crankcase.

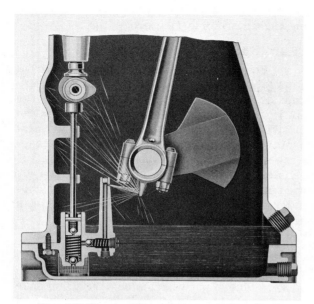

Fig. 5-9 Ejection Pump as it is used on Some Wisconsin Engines.

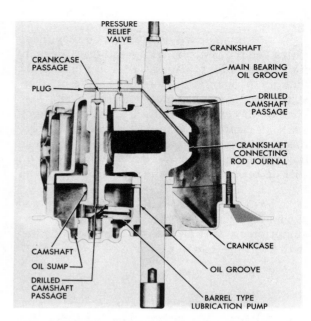

Fig. 5-11 Lubrication Oil Flow of Barrel Type Lubrication System used on Some Lauson Engines.

The Barrel-Type Pump is also driven by an eccentric on the camshaft. The camshaft is hollow and extends to the pump of the vertical crankshaft engine. As the pump plunger is pulled out on intake, an intake port in the camshaft lines up, allowing the pump body to fill. When the plunger is forced into the pump body on discharge, the discharge ports in the camshaft line up, allowing the oil to be forced to the main bearing and to the crankshaft connecting rod journal. Small drilled passages are used to channel the oil. Oil is also splashed onto other crankcase parts.

Full Pressure Lubrication is found on many engines, especially the larger of the small engines and on automobile engines. Lubricating oil is pumped to all main, connecting, and camshaft bearings through small passages drilled in these engine parts. Oil is also delivered to tappets, timing gears, etc. under pressure. The pump used is usually a positive displacement gear type. It is also common to use a splash system in conjunction with full pressure systems.

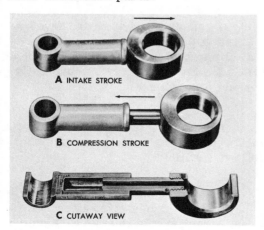

Fig. 5-10 Barrel Type Lubrication Pump used on Some Lauson Engines.

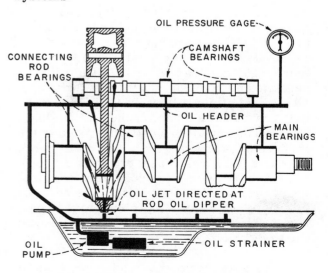

Fig. 5-12 Full Pressure Lubrication

LUBRICATING CYLINDER WALLS

Oil is splashed or sprayed onto cylinder walls and the piston rings spread the oil evenly for proper lubrication. Piston rings must function properly to avoid excessive oil consumption. The rings must exert an even pressure on the cylinder walls and provide a good seal. If piston ring grooves are too large, the piston rings will begin a pumping action as the piston moves up and down, and excess oil will be brought into the combustion chamber.

Worn cylinder walls, worn pistons, and worn rings can all contribute to high oil consumption as well as loss of compression and the resulting loss of power.

BLOW-BY

Blow-by, the escape of combustion gases from the combustion chamber to the crankcase, occurs when piston rings are worn or too loose in their grooves. Carbon and soot from burned fuel are forced into the crankcase by the rings. Much of the carbon is deposited around the rings, hindering their operation further.

Another damaging result of worn rings can be crankcase dilution. If raw, unburned gasoline is in the combustion chamber it can leak by the piston rings and into the crankcase, diluting the crankcase oil and, thereby, reducing the oil's lubricating properties.

CRANKCASE BREATHERS

Four-stroke cycle engines do not have completely airtight, oiltight crankcases. Engine crankcases must breathe. Without the ability to breathe, pressures build up in the crankcase and may rupture oil seals or allow contaminants to remain in the crankcase. Pressure buildup may be caused by the expansion of the air as the engine heats up, by the action of the piston coming down the cylinder and by the blow-by of combustion gases along cylinder walls.

Most single-cylinder engines use breathers that allow air to leave but not reenter a reed-type check valve or a ball-type check valve. These breathers place the crankcase under a slight vacuum and are called closed breathers.

Open breathers allow the engine to breathe freely in and out and are usually equipped with air filters. They are located where the splashing of oil is not a problem. Open breathers are frequently incorporated with the valve access cover. If an engine with this type of breather is tipped on its side, oil may run out through it. Some breathers are vented to the atmosphere while others are vented back through the carburetor.

LUBRICATION OF TWO-CYCLE ENGINES

Lubrication of the two-cycle engine is quite different from the four-cycle engine. Since the fuel mixture must travel through the crankcase, a reservoir of oil cannot be stored there. The lubricating oil is mixed with the gasoline and then put into the gas tank. The lubricating oil for all crankcase parts enters the crankcase as a part of the fuel mixture. Millions of tiny oil droplets suspended in the mixture of gasoline and air settle onto the moving parts in the crankcase, providing lubrication. Oil droplets, being relatively large and heavy, quickly drop out of suspension. Of course, much oil is carried on into the combustion chamber where it is burned along with the gasoline and air.

On a two-cycle engine, oil must be mixed with the gasoline. The engine is not operated on straight gasoline. If it were, the heat of friction would burn up the engine in a short time. It should be pointed out that some two-cycle engine manufacturers are developing and marketing engines with oil metering devices that eliminate the need of premixing the oil and gasoline. Mixing is done automatically in the correct proportions.

In preparing the mixture of gasoline and oil for most two-cycle engines, observe the following rules:

1. Mix a good grade of regular gasoline and oil in a separate container. Do not mix in the gas tank unless it is a remote tank such as is found on many outboards.

2. Pour the oil into the gasoline to insure good mixing and shake the container vigorously. If poorly mixed, the oil will settle to the bottom of the tank, causing hard starting.

3. Strain fuel with a fine mesh strainer as you pour it into the tank to prevent any moisture from entering the tank.

4. Use the oil that is specified for your particular two-cycle engine. Most manufacturers specify their own private brand. In an emergency, other oils may be used, usually SAE 30SB (formerly MM) or SAE 30 SD (formerly MS) nondetergent oil. The manufacturer's brand, however, provides the best lubrication with a minimum of deposit formation.

5. Mix gasoline and oil in the proportions recommended by the engine manufacturer. One common propor-

tion is three-fourths pint of oil to one gallon of gasoline when breaking in a new engine, and one-half pint of oil to one gallon of gasoline for normal use.

ADDITIONAL LUBRICATION POINTS

Whether the engine has two-cycle or four-cycle lubrication it should be remembered that there may well be other lubrication to consider besides the crankcase area. On an outboard engine do not neglect the lower unit which needs a special gear lubricant at several points. If an engine is powering an implement or other machinery there may be transmissions, gear boxes, chains, axles, wheels, shafts, linkages, etc. that need periodic lubrication. Lubricate these additional parts with the oil or lubricant recommended by the manufacturer.

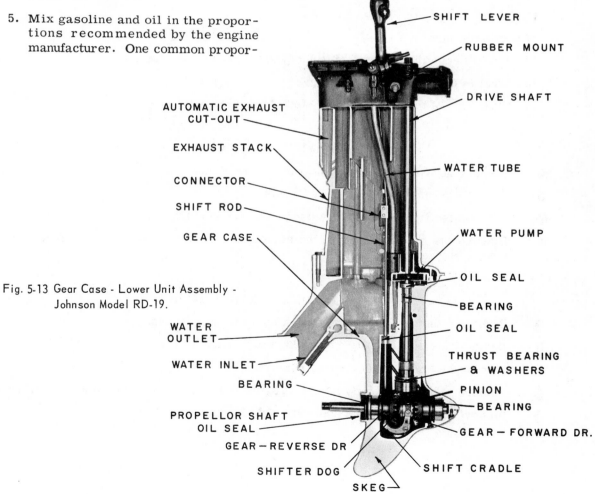

Fig. 5-13 Gear Case - Lower Unit Assembly - Johnson Model RD-19.

GENERAL STUDY QUESTIONS

1. What is the main job of a lubricant?

2. Explain how the lubricating oil also:

   a. Seals power
   b. Helps to dissipate heat
   c. Keeps the engine clean
   d. Cushions bearing loads
   e. Protects against rusting

3. What are the three types of friction bearings?

4. What is an anti-friction bearing?

5. What does an oil's SAE number refer to?

6. What is a multi-grade oil?

7. Explain how a detergent oil works.

8. What are the quality designations of oil?

9. List five common four-cycle engine lubrication systems.

10. How are cylinder walls lubricated?

11. Explain blow-by.

12. Briefly explain how two-cycle lubrication is accomplished.

13. What is one common proportion of oil to gasoline?

14. What kind of oil should be used for two-cycle engines?

CLASS DISCUSSION TOPICS

● Discuss how friction causes heat and wear.

● Discuss the jobs a lubricant performs.

● Discuss the types of bearings used on a " demonstration" engine.

● Discuss the good features of the various lubrication systems; also discuss the system's weak points, if any.

CLASS DEMONSTRATION TOPICS

▶ Illustrate the theory of lubrication using round pencils and a book.

▶ Demonstrate oil viscosity by showing and pouring several different weights of oil.

▶ Remove and inspect oil control rings.

▶ Remove an ejection pump from an engine and demonstrate the pump's operation.

▶ Remove and inspect the dipper on a splash lubrication system.

Unit 6

COOLING SYSTEMS

In internal combustion engines, the temperature of combustion often reaches over 4000° F., a temperature well beyond the melting point of the engine parts. This intense heat cannot be allowed to build up. A carefully engineered cooling system is, therefore, a part of every engine. A cooling system must maintain a good engine operating temperature without allowing destructive heat to build up and cause engine part failure.

The cooling system does not, of course, have to dispose of all the heat produced by combustion. A good portion of the heat energy is converted into mechanical energy by the engine; the more, the better for engine efficiency. Some heat is lost in the form of hot exhaust gases. The cooling system, however, must dissipate about one-third of the heat energy caused by combustion.

Engines are either air-cooled or water-cooled; both systems are in common use. Generally, air-cooled engines are used to power portable machinery, lawnmowers, garden tractors, chain saws, etc. As a rule, the air-cooled system is lighter in weight and simpler; hence, its popularity for portable equipment.

The water-cooled engine is often used for permanent installation or stationary power plants. Most automobile engines are water-cooled, as are all outboard motors.

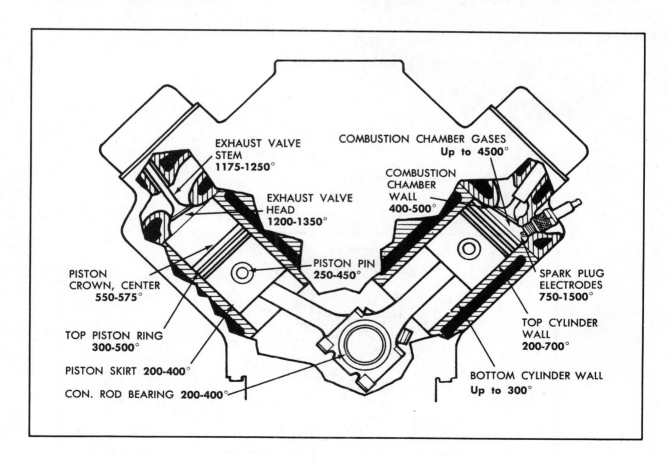

Fig. 6-1 Approximate Engine Operating Temperatures.

AIR-COOLING SYSTEM

The air-cooling system consists of (1) heat radiating fins, (2) flywheel blower, and (3) shrouds for channeling the air.

Heat radiating fins are located on the cylinder head and cylinder because the greatest concentration of heat is in this area. The fins increase the heat radiating surface of these parts allowing the heat to be carried away more quickly.

The flywheel blower consists of air vanes cast as a part of the flywheel. As the flywheel revolves, these vanes blow cool air across the fins, carrying away the heated air and replacing it with cool air.

Fig. 6-3 The Air Shroud is an Important Part of the Air-cooled Engine.

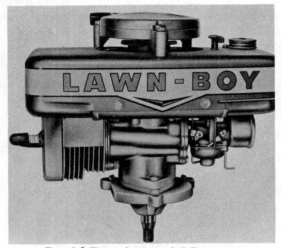

Fig. 6-2 Typical Air-cooled Engine.

The shrouds direct the path of the cool air to the areas that demand cooling. The shroud may look like a decorative cover serving no real purpose but this is not true. Shrouds must be in place if the cooling system is to operate at its maximum efficiency.

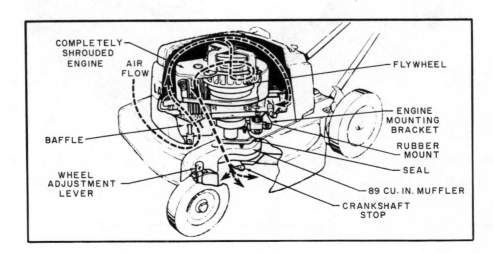

Fig. 6-4 The Path of Air Flow on Certain Lawn-Boy Engines.

CARE OF THE AIR-COOLING SYSTEM

The air-cooling system is almost foolproof, but not infallible. Several points should be considered. Heat radiating fins are thin and often fragile, especially on aluminum engines. If, through carelessness, they are broken off, a part of the cooling system is gone. Besides losing some cooling capacity, hot spots may develop, warping the damaged area. Also, it is easy for grass, and oil to accumulate between the fins. As an accumulation builds up, the cooling system's efficiency goes down. Keep the radiating fins clean.

The flywheel vanes should not be chipped or broken. Besides reducing the cooling capacity of the engine, such damage may destroy the balance of the flywheel. An unbalanced flywheel will cause vibration and more than the normal amount of wear on engine parts.

WATER COOLING SYSTEM

Many small engines are water-cooled as are most larger engines. Such engines have an enclosed water jacket around the cylinder walls and cylinder head. Cool water is circulated through this jacket, picking up the heat and carrying it away.

The basic parts of a water-cooling system such as is commonly found on stationary small gas engines or an automobile engine are: radiator, fan, thermostat, water pump, hoses, and water jacket.

The water pump circulates the water throughout the entire cooling system. The hot water from the combustion area is carried from the engine proper to the radiator. In the radiator, many small tubes and radiating fins dissipate the heat into the atmosphere. A fan blows cooling air over the radiating fins. From the bottom of the radiator the cool water is returned to the engine.

Engines are designed to operate with a water temperature of between 160° to 180° F. To maintain the correct water temperature a thermostat is used in the cooling system. When the temperature is below the thermostat setting, the thermostat remains closed and the cooling water circulates only through the engine. However, as the heat builds up to the thermostat setting, the thermostat opens and the cooling water moves throughout the entire system.

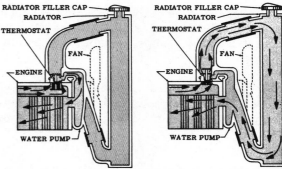

Fig. 6-6 Thermostat Closed: Water Recirculated through Engine Only. Thermostat Open: Water Circulated through Both Engine and Radiator.

WATER COOLING THE OUTBOARD ENGINE

Cooling the outboard engine with water is a simpler process because there is an inexhaustible supply of cool water present where the engine operates. Outboards pump water from the source through the engine's water jacket and then discharge the water back into the source.

Fig. 6-5 Cross-section Showing Water Passages in Head and Block.

The water pump on outboards is located in the lower unit. It is driven by the main driveshaft or the propeller shaft. The cool water is pumped up copper tube passages to the water jacket. After the cooling water picks up heat, it is discharged into the exhaust area of the lower unit and out of the engine.

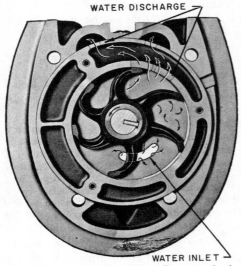

Fig. 6-7 An Impeller Water Pump with the Cover Removed to Show the Impeller.

Several types of water pumps are used on outboard motors: Plunger Type Pumps, Eccentric Rotor Pumps, Impeller Pumps, and others.

Many outboard motors, especially the recent, large horsepower models, are equipped with a thermostatically controlled cooling

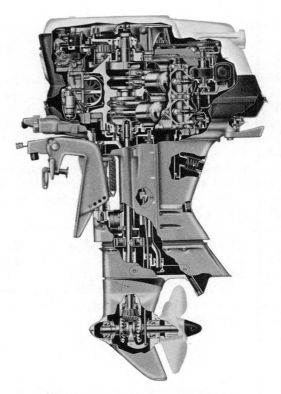

Fig. 6-9 Cutaway of an Outboard Motor.

system. The temperature of the water circulating through the water jacket is maintained at about 150° F.

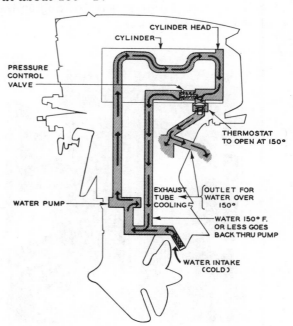

Fig. 6-10 Thermostat Controlled Cooling System used on Outboard Engines.

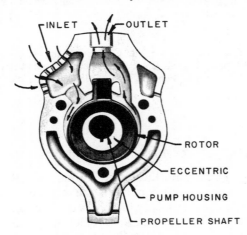

Fig. 6-8 An Eccentric Rotor Water Pump with the Cover Removed to Show the Eccentric and Rotor.

GENERAL STUDY QUESTIONS

1. How high can the temperature of combustion reach?

2. Look up the melting points of aluminum and iron.

3. Does the cooling system remove all the heat of combustion? Explain.

4. Why are air-cooling systems often used for "portable" equipment?

5. What are the main parts of the air-cooling system?

6. Why are cylinders and cylinder heads equipped with fins instead of being cast with a smooth surface?

7. What are the main parts of the water-cooling system?

8. What is the function of the thermostat?

9. What are several types of water pumps used on outboard motors?

CLASS DISCUSSION TOPICS

● Discuss how heat can damage engine parts.

● Discuss the path of heat flow from the inside of the engine to the outside.

● Discuss the advantage of the air-cooled engine.

● Discuss the advantages of the water-cooled engine.

● Discuss why it is best for the engine to operate at a constant temperature.

CLASS DEMONSTRATION TOPICS

▶ Trace the air flow through an air-cooled engine.

▶ Trace the water flow through an outboard engine.

▶ Disassemble various types of water pumps; show their operation and construction.

▶ Show how the air-cooling system can be damaged.

IGNITION SYSTEMS

Small gasoline engines normally use a magneto for supplying the ignition spark. A magneto is a self-contained unit that produces the spark for ignition; no outside source of electricity is necessary. It is a simple and very reliable ignition system. Since most small gas engines do not have electric starters, lighting systems, radios and other electrical accessories, a storage battery is not necessary. The magneto is, therefore, ideally suited for the small gasoline engine.

The basic parts of the magneto ignition system are: (1) permanent magnets, (2) high tension coil (primary and secondary), (3) laminated iron core, (4) breaker points, (5) breaker cam, (6) condenser, (7) spark plug cable, and (8) spark plug. Before trying to understand how these parts work together, it will be well to review some essentials of electricity and magnetism and also to study the construction and function of each individual part.

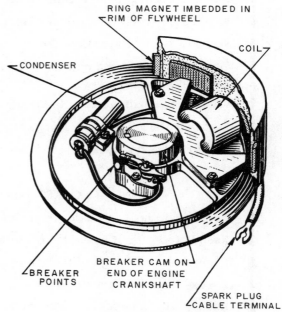

Fig. 7-1 Flywheel Magneto.

ELECTRON THEORY

All matter is composed of atoms and, of course, these atoms are infinitesimally small. The atom, itself, is composed of electrons, protons, and neutrons. The number and arrangement of these particles determines the type of atom: hydrogen, oxygen, carbon, iron, lead, copper, or any other element. Weight, color, density, and all other characteristics of an element are determined by the structure of the atom. Electrons from an atom of copper would be the same as electrons from any other element.

The electron is a very light particle that spins around the center of the atom. Electrons move in an orbit. The number of electrons orbiting around the center or nucleus of the atom varies from element to element. The electron has a negative (-) electrical charge.

The proton is a very large and heavy particle in relationship to the electron. One or more protons will form the center or nucleus

of the atom. The proton has a positive (+) electrical charge.

The neutron consists of an electron and proton bound tightly together. Neutrons are located near the center of the atom. The neutron is electrically neutral; it has no electrical charge.

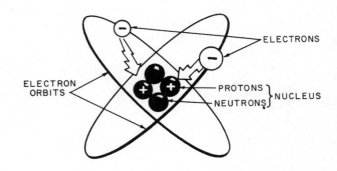

Fig. 7-2 Atomic Structure: Electron, Proton, and Neutron.

Atoms are normally electrically neutral, that is, the number of electrons and protons are the same, cancelling out each other's electrical force. Atoms "stay together" because unlike electrical charges attract each other. The electrical force of the protons holds the electrons in their orbits. Like electrical charges repel each other so negatively charged electrons will not collide with each other.

In most materials it is very difficult, if not impossible, for electrons to leave their orbit around the atom. Materials of this type are called nonconductors of electricity, or insulators. Some typical insulating materials are glass, mica, rubber, paper, etc. Electricity will not flow through these materials.

In order to have electric current, electrons must move from atom to atom. Insulators will not allow this electron movement.

However, there are quite a few substances in which it is relatively easy for an electron to jump out of its orbit and begin to orbit in an adjoining or nearby atom. Substances which permit this movement of electrons are called conductors of electricity. Everyone is familiar with such typical examples as copper, aluminum and silver.

Electron flow in a conductor takes place when there is a difference in electrical potential and there is a complete circuit or path for electron flow. Another way of stating this is that the source of electricity is short of electrons; it is positively charged and since unlike charges attract each other, electrons, being negatively charged, will move toward the positive source.

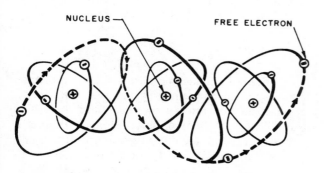

Fig. 7-3 Current: Flow of Electrons Within a Conductor.

A source of electricity can be produced or seen in three basic forms: (1) mechanical, (2) chemical, (3) static. Electricity is produced mechanically in the electrical generator which is commonly lashed to water power or steam turbines. The electricity we use in our homes and factories is produced mechanically. In the magneto, mechanical energy is used to rotate the permanent magnet. Electricity produced by chemical action is seen in the storage battery and dry cell. Static electricity can be seen in nature when lightning strikes. The lightning occurs when the air insulation breaks down and electrons are in a positive area. The lightning may be between clouds, from cloud to earth, or from earth to cloud.

UNITS OF ELECTRICAL MEASUREMENT

There are three basic units of electrical measurement:

1. Rate of electron flow - amperes.

2. Force or pressure causing electron flow - volts.

3. Resistance to electron flow - ohms.

The ampere is the measurement of electrical current, the number of electrons flowing past a given point in a given length of time. If you could stand at a point on a wire and count the electrons passing by in one second and you counted 6,250,000,000,000,000,000 you would have counted one ampere of current. To help visualize amperage, think of water flowing in a pipe. A small pipe might deliver two gallons of water a minute. A larger pipe might deliver five gallons of water a minute. Electric wires are generally the same; larger wires can handle more amperage or electron flow than smaller wires.

The volt is the measurement of electrical pressure or the difference in electrical potential that causes electron flow in an electrical circuit. The energy source is short of electrons and the electrons in the circuit want to go to the source. The pressure to satisfy the source is called voltage. Voltage might be compared to the pressure that water in a high tank places on the pipe located at the street level. The higher the water pressure, the faster the water flow from a pipe below. Likewise, a higher voltage tends to cause greater flow of electrons.

The <u>ohm</u> is the unit of electrical resistance. Every substance puts up some resistance to the movement of electrons. Insulators such as porcelain, oils, mica, glass, etc. put up a tremendous resistance to electron flow. Conductors such as copper, aluminum, and silver put up very little resistance to electron flow. Even though conductors readily permit the flow of electric current they do tend to put up some resistance. In the water pipe example, this resistance might be seen as the surface drag by the sides of the pipe, or scale and rust in the pipe. Using a larger pipe, or, electrically, a larger wire, is one way of reducing resistance.

OHM'S LAW

In every example of electricity flowing in an electrical circuit, amperes, volts, and ohms each play their part; they are related to each other. This relationship is stated in Ohm's Law.

$$\text{Amperes (rate)} = \frac{\text{volts (potential)}}{\text{ohms (resistance)}}$$

The formula is usually abbreviated to:

$$I = \frac{E}{R}$$

For example, if the voltage were 6, and the resistance 12 ohms, calculate the current.

$$I = \frac{E}{R} \quad I = \frac{6}{12} \quad I = .5 \text{ amperes}$$

Of course, the formula can be written to find the resistance or the voltage.

$$R = \frac{E}{I} \quad E = IR$$

MAGNETISM

No doubt everyone has played with a magnet at sometime, watched it pick up steel objects and watched it attract or repel another magnet. These effects are curious and still not entirely explained, but scientists generally agree on the molecular theory of magnetism. Molecules are the smallest divisions of substance that are still recognizable as that substance. Several different atoms may make up one molecule. For example, a molecule of iron oxide will contain atoms of iron and oxygen. In many substances the atoms in the molecules are

more positive at one spot and more negative at another spot. This is termed a north pole and a south pole. Usually the poles of adjoining molecules are arranged in a random pattern and there is no magnetic force since their effects cancel one another. However, in some substances, such as iron, nickle, and cobalt, the molecules are able to align themselves so that all north poles point in one direction and all south poles point in the opposite direction. The small magnetic forces of many tiny molecules combine to make a noticeable magnetic force. In magnets, like poles repel each other, unlike poles attract each other, just as like and unlike electrical charges react. Electricity and magnetism are very closely tied together.

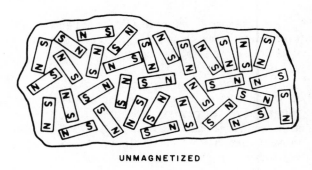

UNMAGNETIZED

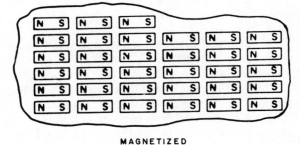

MAGNETIZED

Fig. 7-4 In an Unmagnetized Bar the Molecules Are in a Random Pattern. In a Magnetized Bar the Molecules Align Their Atomic Poles.

Some substances can retain their molecular alignment permanently and are, therefore, classed as permanent magnets. Hard steel has this ability. A piece of soft iron such as a nail can attain the molecular alignment of a magnet only when it is in a magnetic field. As soon as soft iron is removed from the magnetic field, its molecules disarrange themselves and the magnetism is lost.

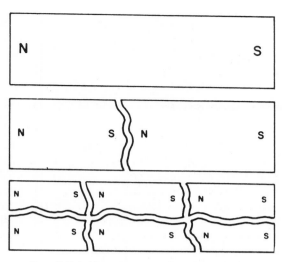

Fig. 7-5 A Magnetic Field Surrounds Every Magnet. Like Poles Repel Each Other; Unlike Poles Attract Each Other.

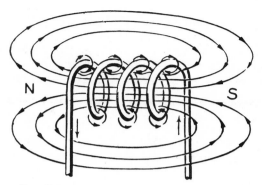

Fig. 7-7 When Electricity Flows Through a Coil of Wire, a Magnetic Field is Set Up around the Coil.

More than 100 years ago Michael Faraday discovered that magnetism could produce electricity. Magnetos used on gasoline engines use this discovery: magnetism producing electricity. Faraday found that if a magnet is moved past a wire, electrical current will start through the wire. If the magnet is stopped near the wire, the current will stop. Electricity will flow only when the magnetic field or magnetic lines of force are being cut by the wire.

The principle of the transformer and induced voltage is also used in the magneto. In a transformer there is a primary coil and a secondary coil wound on top of the primary; the two are insulated from each other. These coils are wound on a soft iron core. When alternating current passes through the primary coil there is an alternating magnetic field set up in the iron core. The magnetic lines of force cut the secondary coil and induce an alternating voltage within the coil. The voltage produced depends on the ratio of windings in the primary coil and secondary coil. If there are more windings in the secondary than the primary, the secondary voltage will be higher, a step-up transformer. If there are more windings in the primary than the secondary, the secondary voltage will be smaller, a step-down transformer. Although the magneto does not operate on alternating current, it does use the principle of the step-up transformer.

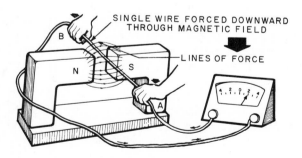

Fig. 7-6 Current Flows in the Wire As It Moves Down Through the Magnetic Field.

Another principle that the magneto uses is that when electrons flow through a coil of wire, a magnetic field is set up around the coil. The coil itself becomes a magnet. Therefore, when electrons flow through the coils in a magneto, a magnetic field is set up.

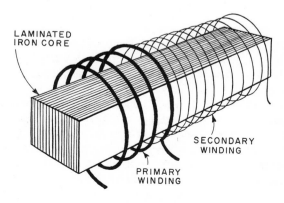

Fig. 7-8 A Simple Transformer.

In a small gasoline engine magneto, there is a magnetic field which induces current in the primary coil, thus setting up a magnetic field around both the primary and secondary coils. At the point of maximum current, the circuit is broken in the primary. Electrons can no longer flow; therefore, the magnetic field collapses. This rapidly collapsing magnetic field induces a very high voltage, igniting the fuel mixture.

BASIC MAGNETO PARTS

Before studying in detail how the magneto operates, it would be well to examine the basic parts.

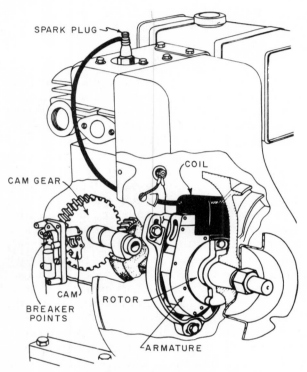

Fig. 7-9 The Basic Parts of a Briggs-Stratton Magneto Ignition System.

PERMANENT MAGNETS

Permanent magnets are usually made of an alloy called Alnico, a combination of aluminum, nickle and cobalt. This magnet is quite strong and will retain its magnetism for a very long time. On a flywheel magneto the magnet is cast into the flywheel and cannot be removed. The other magneto parts often are mounted on a fixed plate underneath the flywheel. The magnet, therefore, revolves around the other parts of the magneto. Sometimes the other magneto parts are mounted near the outside rim of the flywheel and the permanent magnets pass by each revolution.

Rotor type permanent magnets are also in use. With this type of construction the permanent magnet rotor may be mounted on the end of the crankshaft or the rotor may be geared to the crankshaft. The rotor lies "within" the other magneto parts. Care should be taken not to drop or pound the magnet as this will cause it to lose some of its magnetism.

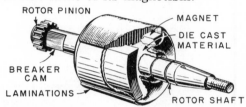

Fig. 7-10 Rotor Type Permanent Magneto.

HIGH TENSION COIL
(PRIMARY AND SECONDARY WINDINGS)

The primary winding of wire consists of about 200 turns of a heavy wire (about 18 gage) wrapped around a laminated iron core. The primary coil is in the electrical circuit containing the breaker points and condenser. When a magnet is brought near this coil and iron core, magnetic lines of force cut the coil and electrical current is produced. Normally this current flows from the coil, through the closed breaker points and into ground, a complete circuit.

The secondary winding of wire consists of about 20,000 turns of a very fine wire wrapped around the primary coil. The secondary coil is in the electrical circuit containing the spark plug. When current flows in the primary, a magnetic field gradually expands around both coils, but voltages produced in the secondary are quite small. However, when the breaker points open, the circuit is broken, electricity stops flowing and the magnetic field suddenly collapses. The suddenly collapsing magnetic field induces very high voltage in the secondary coil, enough to jump the spark gap in the spark plug.

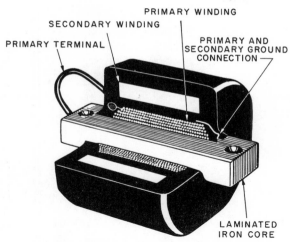

Fig. 7-11 Construction of A High Tension Magneto Ignition Coil.

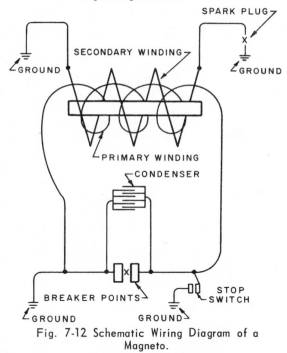

Fig. 7-12 Schematic Wiring Diagram of a Magneto.

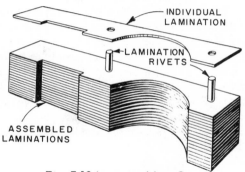

Fig. 7-13 Laminated Iron Core.

LAMINATED IRON CORE

The laminated iron core is made of many strips of soft iron fastened tightly together. The soft iron core helps to strengthen the magnetic field around the primary and secondary coils, but the core will not retain the magnetism and become permanently magnetized. The purpose of using many strips instead of a solid core is to reduce eddy currents which create heat in the core. The general shape of the laminated core may vary from magneto to magneto but its function remains the same.

BREAKER POINTS

The breaker points are made of tungsten and mounted on brackets. The two points are normally closed or touching each other, providing a path for electron flow. However, just before the spark is desired, the breaker points are opened, breaking the electrical circuit. Breaker points open and close anywhere from 800 times per minute to 4500 or more times per minute, depending on the engine speed.

When the breaker points are open they are usually separated by .020 of an inch; the exact opening varies from magneto to magneto. This separation is critical to the function of the magneto; therefore, the breaker point gap must be adjusted correctly.

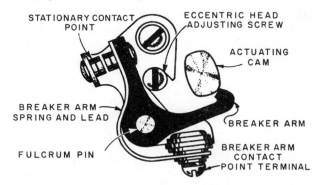

Fig. 7-14 Breaker Point Assembly.

BREAKER CAM

The breaker cam actuates the breaker points. One bracket of the breaker assembly rides on this cam. As the cam rotates, it opens and closes the breaker points.

On most two-cycle engines this cam is mounted on the crankshaft and opens the normally closed breaker points once each revolution. On most four-cycle engines, it is mounted to operate from the camshaft which is turning at one-half crankshaft speed. By mounting the breaker points here, they open once every two revolutions of the crankshaft, thereby providing a spark for the power stroke but none for the exhaust stroke.

The shape of the cam depends on the number of cylinders and the design of the magneto.

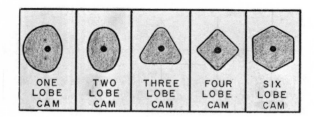

Fig. 7-15 Common Breaker Cam Shapes.

CONDENSER

The condenser acts as an electrical storage tank in the primary circuit. When the breaker points open quickly, the electrons tend to keep flowing and, if no condenser were present, a spark might actually jump across the breaker points. If this happened, the breaker points would soon burn up, not to mention weakening effect on the voltage produced by the magneto secondary coil. The condenser provides an electrical storage tank for this last surge

Fig. 7-17 Typical Condensers.

of electron flow. The condenser can be easily located because it usually looks like a miniature tin can.

SPARK PLUG CABLE

The spark plug cable connects the secondary coil and the spark plug, providing a path for the high-tension voltage. This part is often referred to as the high-tension lead.

SPARK PLUG

The spark plug is a vital part of the ignition system. In this part the resulting work of the magneto parts is seen. Basically, the spark plug consists of a shell, ceramic insulator, center electrode and ground electrode. The two electrodes are separated by a gap of about .030 of an inch. The path of electricity is down the center electrode, across the air gap, to the ground electrode. Of course, the voltage must jump the air gap. When a high-tension voltage of about 20,000 volts is reached, a spark will jump between the electrodes. This spark ignites the fuel mixture within the combustion chamber.

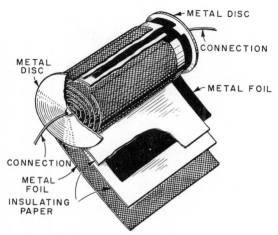

Fig. 7-16 Cutaway Showing the Construction of a Condenser.

THE COMPLETE MAGNETO CYCLE

Now let us "walk through" one revolution of the crankshaft and observe the flywheel magneto operation. When the permanent magnet is far away from the high-tension coil, it has no effect on the coil. But as the permanent magnet comes closer and closer, the primary coil "feels" the increasing magnetic field; the coil is being cut by magnetic lines of force, and, therefore, electrons flow in the primary coil. This current passes through the breaker points and into the ground. When the permanent magnet is just about opposite the high-tension coil the magnetic field around both coils is reaching its peak. Also, the piston is reaching the top of its stroke, compressing the fuel mixture.

The position of the permanent magnet is now causing the polarity of the laminated iron core and the coils to reverse direction. The reversing of direction is momentarily choked or held back by the coil. Then the breaker points open, interrupting the current flow and allowing the reversal and subsequent rapid collapse of the intense magnetic field that had been built up around the primary and secondary coils. Magnetic lines of force are cutting coils very rapidly, and high voltages are induced in the coils. In the secondary coil the voltage may reach 18,000 to 20,000 volts, enough to jump across the spark gap in the spark plug. This spark ignites the fuel mixture and the piston is forced down the cylinder.

SPARK ADVANCE

When fuel burns in the combustion chamber, it does not explode and exert all of its power instantaneously. A very short period of time is required for the fuel to ignite and reach its full power. True, this time is very short but in an engine operating at high speeds this small time lag is important.

For example, if we waited until the piston reached dead center before igniting the mixture, the piston would already be started back down the cylinder before the full force of the burning fuel is reached. This results in loss of power.

Therefore, it is necessary to ignite the fuel slightly before the piston reaches the top of its stroke to realize the full force of combustion. Causing the spark to occur earlier in the engine cycle is called spark advance.

The spark is advanced more and more as the engine speed increases, because there is less time for combustion to take place. For example, in a four-cycle engine operating at 2,000 r.p.m., each power stroke takes about 1/64 of a second, but if the speed is increased to 4,000 r.p.m., each power stroke will take only 1/128 of a second. If the speed is doubled, there is only one-half as much time for combustion to take place. Also, high speeds give the engine higher compression and more explosive mixture. At high speeds the spark jumps the spark gap before the piston reaches the top of its stroke.

At slow speeds the spark is retarded and occurs later in the cycle, slightly before the piston reaches top dead center or sometimes at top dead center. At slow speeds there is more time for combustion to take place. Also, compression is lower and not as much explosive mixture is drawn into the combustion chamber.

Regulating spark advance is done by controlling the time that the breaker points open. Advancing the spark can be done automatically or manually; both methods are commonly used on small gasoline engines.

MANUAL SPARK ADVANCE

Manual spark advance is usually accomplished by loosening the breaker point assembly and rotating it slightly to an advanced position. It is then locked in its new position.

Many outboard motors have a type of manual spark advance although it is generally referred to by manufacturers as a "spark-gas" synchronization system. The magneto plate, on which the breaker points are mounted, is located underneath the flywheel. This plate can be rotated through about 25°. The breaker cam is securely mounted on the crankshaft; its relative position cannot be changed. Spark advance is obtained by moving the magneto plate so the breaker points will be opened earlier in the cycle. The throttle and magneto plates are linked together so that opening the throttle also advances the spark. At full throttle the spark is fully advanced. At idling speeds the throttle is closed and the magneto plate is positioned for minimum spark advance.

AUTOMATIC SPARK ADVANCE

Automatic spark advance is usually accomplished by a centrifugal mechanism which is capable of changing the relative position of the breaker cam and the breaker points. Here the breaker cam can be rotated through a small distance on its shaft, about 25°. A spring holds the cam in the retarded position at idling and slow speeds but as the engine speed increases, centrifugal force throws the mechanism's hinged weights outward. This outward motion overcomes the spring's tension and this motion is used to rotate the cam to a more advanced position. At full speed the cam has been rotated its full limit for maximum spark advance.

Some engines, especially larger types, also have a vacuum advance mechanism working with a centrifugal mechanism to provide more accurate spark advance, especially at slow speeds. Automatic spark advance can also be used to move the breaker point assembly to secure the proper advance.

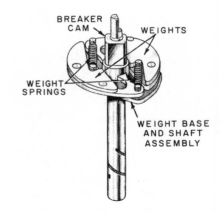

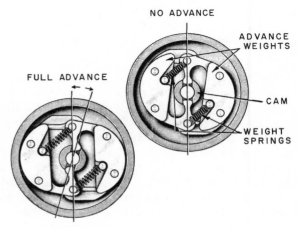

Fig. 7-19 Automatic Spark Advance Mechanism Changes the Breaker Cam's Position in Relation to the Points.

IMPULSE COUPLING

When an engine is started by hand, it is turned over slowly. Since the voltages produced by the magneto depend upon the speed that magnetic lines of force cut the primary coil, the voltages produced for starting can be quite low, resulting in a weak spark.

Some engines are equipped with an impulse coupling device to supply higher voltages and a hotter spark for slow starting speeds. Using a tight spring and retractable pawls, the rotation of the magneto's magnetic rotor can be stopped for all but the last few degrees of its revolution. During the last few degrees of the revolution the pawl retracts, allowing the spring to snap the magnetic rotor past the firing position at a very high speed. A high voltage can be produced because magnetic lines of force are cutting the primary coil rapidly.

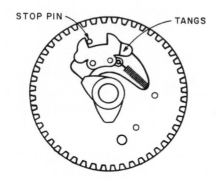

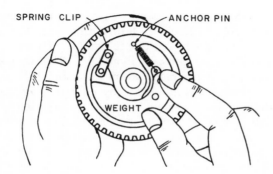

Fig. 7-18 Spark Advance Mechanism with Breaker Cam and Weight Mounted on the Camshaft Gear.

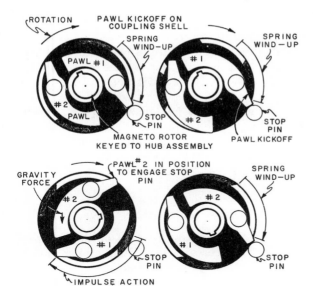

Fig. 7-20 Sequence of Operation for Impulse Coupling for 180° Spark Magneto.

MULTI-CYLINDER ENGINES

Magnetos are often used for multi-cylinder engines. However, having more than one cylinder to supply with a spark does present a problem. Two solutions to this problem are in common use today.

Two-, three- and four-cylinder engines can have magneto ignition by simply installing a separate magneto for each cylinder. This method is commonly used with outboard engines using a flywheel magneto. Instead of mounting just one magneto on the armature plate under the flywheel, a magneto is installed for each cylinder. The breaker points of each magneto

ride on the common cam. In two-cylinder engines the magnetos would be 180° apart, three-cylinder engines 120° apart, and four-cylinder engines 90° apart. Each magneto functions separately to supply its cylinder with a spark.

Another solution is to use a distributor and two-pole or four-pole magnetic rotor. Only one complete magneto is used even though the engine may have two or four cylinders. In the case of a two-pole rotor used on a two-cylinder engine, the magneto can produce two sparks every revolution of the magnetic rotor, one every 180° rotation. A two-lobed cam is used to open the breaker points twice each rotor revolution. The distributor rotor channels the high-tension voltage to the correct spark plug.

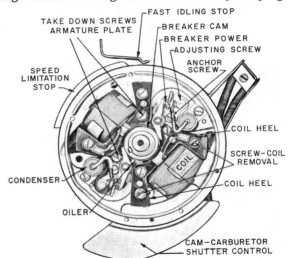

Fig. 7-21 Two Complete Magnetos are Installed on this Armature Plate.

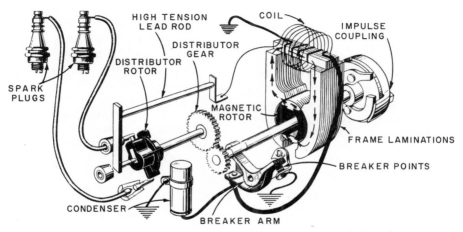

Fig. 7-22 Diagram of Rotating Magnet Magneto with Jump-Spark Distributor.

ENGINE TIMING

The spark must jump the spark gap at exactly the right time, just before the piston reaches top dead center. In many engines, the breaker cam is driven by the camshaft, therefore, the gear on the crankshaft and the camshaft gear must be assembled correctly. These two gears are marked in some manner, usually punch marks, so the repairman can easily make the correct assembly. Incorrect alignment of the gears may cause poor operation or the engine may not operate at all.

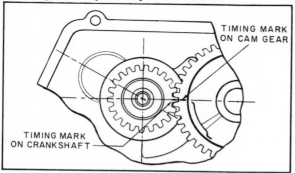

Fig. 7-23 Timing Marks on Crankshaft Gear and Cam Gear Must Be Aligned Correctly.

SOLID STATE IGNITION

Solid state ignition refers to the fact that solid state electronic parts, namely transistors, replace the breaker points. The breaker points in a conventional ignition system can be a source of trouble. Sometimes the systems are called Capacitor Discharge Systems, Breakerless Ignition, or Transistorized Ignition. These systems may be used with just a conventional flywheel magnet as the source of the magnetic force. They may be used with a generator or an alternator battery system.

The main components of the solid state ignition system are a generator or alternator coil, trigger module, ignition coil assembly, and special flywheel with trigger projection. The system has a conventional spark plug and lead.

The trigger module contains transistor diodes which rectify the alternating current, changing it into direct current. Transistors control the flow of electric current by acting somewhat like a valve. Their resistance to flow can be changed by a small current to the transistors. In addition to the diode rectifiers, the trigger module contains a resistor, a sensing coil and magnet, and a silicon-controlled rectifier (SCR). The SCR acts as a switch.

The ignition coil contains primary and secondary windings similar to those of a conventional magneto coil plus a condenser or capacitor. Electrically speaking, capacitors and condensers are basically the same.

The operation of the solid state ignition system follows this cycle:

1. The rotating magnet on the flywheel sets up an alternating current in the alternator or generator coil. This alternating current is rectified into direct current and stored in the capacitor. The diode rectifiers permit flow in one direction only so the capacitor cannot discharge back through the diodes.

2. The magnet group passes the trigger coil, setting up a small current which triggers or gates the SCR. This gating makes the SCR conductive so that the stored voltage of the capacitor surges from capacitor through the SCR and is applied across the primary winding of the ignition coil and to the negative side of the capacitor. This instantaneous surge of energy sets up a magnetic field in the primary winding of the ignition coil. The magnetic field is also induced around the secondary coil, and voltages sufficient to jump the spark gap at the spark plug are reached.

Solid state ignition offers several advantages, such as automatic retarding of the spark at starting speeds, longer spark plug life, faster voltage rise, and a high energy spark which makes the condition of the plug and its gap less critical.

SPARK PLUG

The spark plug is the part of the ignition system that ignites the fuel mixture. It operates under severe and varying temperature conditions and is a critical part in engine operation.

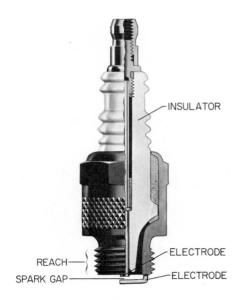

Fig. 7-24 Cutaway of a
Spark Plug.

will quickly burn up. The heat range for spark plugs depends on how fast the heat can be carried away from the electrodes. The path of heat transfer is from the electrodes to the ceramic insulator through the shell and into the cylinder head. The length of the ceramic insulator exposed to the combustion chamber determines the heat range of a spark plug. The longer this insulator, the longer the heat path, and the "hotter" the spark plug's operating temperature. Likewise, the shorter this insulator, the shorter the heat path, and the "colder" the spark plug's operating temperature.

The spark gap or distance between the electrodes must be correctly set. It is within this small space that the spark jumps and combustion begins. The gap must be large enough for sufficient fuel mixture to "get between" the electrodes but not so large as to prevent the spark from jumping across. The spark gap for various plugs ranges from about .020" to .040".

All spark plugs are basically the same but they do differ in these respects: (1) thread size, (2) reach, (3) heat range, (4) spark gap. There are hundreds of different types of spark plugs, each designed for the special requirements of a certain engine.

The shell of the spark plug is threaded so that it can easily be installed or removed from the cylinder head. Various spark plugs have different thread sizes. Some of the more common standard thread sizes are 7/8", 10 mm., 14 mm., and 18 mm.

The reach of the spark plug is the distance between the gasket seat and the bottom of the spark plug shell, or roughly the length of the cut threads. The reach ranges from about 1/4" to 3/4". Each engine must be equipped with a spark plug of the correct reach since reach determines how far the electrodes protrude into the combustion chamber. If the reach is too small, the electrodes will find it difficult to ignite the fuel when the spark jumps. If the reach is too great, the top of the piston may strike the electrodes on its upward stroke.

The heat range of a spark plug is the range of temperature within which the spark plug is designed to operate. If a spark plug operates at too low a temperature, it will quickly foul with oil and carbon. If the spark plug operates at too high a temperature, the electrodes

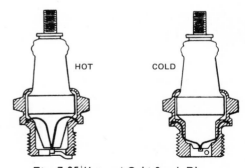

Fig. 7-25 Hot and Cold Spark Plugs.

If the spark plug is removed for cleaning, the gap should be checked and reset according to the manufacturer's specifications. A wire gage should be used.

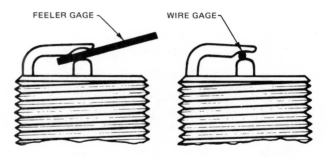

Fig. 7-26 A Plain, Flat Feeler Gage Cannot Accurately Measure the True Width of a Spark Gap.

GENERAL STUDY QUESTIONS

1. What three particles make up the atom?

2. Explain how electrons can move in a conductor.

3. What are the three ways that electricity can be produced or seen?

4. Explain the electrical measurements: (1) amperes, (2) volts, and (3) ohms.

5. State Ohm's Law.

6. Explain how permanent and temporary magnets differ.

7. What happens when a wire is cut by magnetic lines of force?

8. What happens to a coil of wire when electric current flows through it?

9. Explain how a transformer works.

10. In a magneto, what is the source of the magnetic field?

11. What are the two parts of the high-tension coil?

12. What purpose does the laminated iron core serve?

13. What is the function of the breaker points? How are they opened?

14. What is the function of (1) the condenser? (2) the spark plug?

15. What are the two electrical circuits in the magneto?

16. Why must the spark be advanced at high speeds?

17. What is the advantage of impulse coupling?

18. Can magneto ignition be used for multicylinder engines?

19. In what four ways do spark plugs differ?

CLASS DISCUSSION TOPICS

- Discuss atomic structure.
- Discuss how conductors and insulators differ.
- Discuss the molecular theory of magnetism.
- Discuss the advantages of magneto ignition.
- Discuss the complete magneto cycle.

CLASS DEMONSTRATION TOPICS

- Demonstrate a magnetic field around a magnet using a permanent magnet, iron filings, and paper.
- Demonstrate current flow caused by a varying magnetic field using a galvanometer, coil of wire, and permanent magnet.
- Demonstrate a magnetic field around a coil of wire using a d-c source, coil of wire, iron filings, and paper.
- Disassemble a magneto showing the basic parts.
- Demonstrate how the breaker points are opened and show their separation at maximum opening (feeler gage).

ROUTINE CARE AND MAINTENANCE - AND WINTER STORAGE

A gasoline engine represents an investment in money, perhaps only fifty dollars, but possibly several hundred dollars. To safeguard this investment the engine operator must perform certain routine steps in care and maintenance. Routine care and maintenance will insure the longest life possible for engine parts and may save costly repair bills.

Certainly there are persons who provide little or no special care for their engines. For example, they may operate an engine all summer with no regard for its needs, other than gasoline. When fall comes, the engine is simply parked in the corner of the garage until the next spring.

Such engine owners are gamblers; they may get away with their neglect for one year, two years, or three years, but, on the other hand, they may soon have a depreciated, worn out engine that is still quite young. The correct care of the engine will insure many years of useful life with a minimum of repair bills.

The best source of information on engine care is the "Owner's Manual" or "Operator's Instruction Book". These books, supplied with every new engine, are prepared by the engine manufacturer in order to acquaint the engine owner or operator with that particular engine's requirements.

ROUTINE CARE AND MAINTENANCE

Routine care and maintenance is given to an engine during the engine's normal use. It is provided to keep the engine operating at its peak efficiency and to prevent undue wear of engine parts. There are four basic points to consider: (1) oil supply, (2) cooling system, (3) spark plug, and (4) air cleaner.

OIL SUPPLY

On a four-cycle engine, the oil supply in the crankcase should be checked daily or each time the engine is used. If the oil level has dropped below the add mark, fill the crankcase to the proper level.

Most manufacturers recommend that the oil be changed after a short period of time when breaking in a new engine. Depending on the manufacturer, this first oil change should be accomplished from 2 to 20 hours.

After the engine has had its oil changed once, it can go for a longer period of time between oil changes. Lauson engines specify changing the oil every 10 hours of operation; Briggs & Stratton, Whizzer, Kohler, and Clinton engines every 20 hours; Gravely and Wisconsin engines every 50 hours; Onan engines after 100 hours. Each manufacturer's instruction book will give the length of time between oil changes.

Remember, on two-cycle engines, oil changes are not necessary since all lubricating oil is mixed with the gasoline.

COOLING SYSTEM

The most important consideration regarding the cooling system is to keep it clean. On air-cooled engines, clean the radiating fins, flywheel vanes, and shrouds whenever they begin to accumulate grass, dirt, etc. Do not allow deposits to build up, reducing the engine's cooling capacity. How often the cooling system is cleaned depends on the dust conditions under which the engine operates.

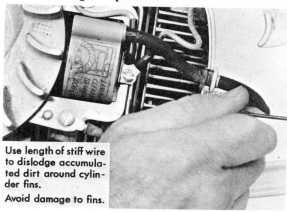

Use length of stiff wire to dislodge accumulated dirt around cylinder fins.

Avoid damage to fins.

Fig. 8-1 Keep Heat Radiating Fins Clean.

On a water-cooled engine, check the water level in the radiator before operating the engine. If it is low, fill it to the correct level.

In the case of an outboard motor, there is little routine care required for the cooling system. However, each time the engine is started the operator should check to see that the engine is "pumping water". This can be detected on most engines by a small amount of water coming from the "telltale holes". If the engine is not pumping water it should be stopped immediately and the source of the trouble corrected. Do not put your hand by the underwater discharge to determine if the engine is pumping water; you may be dangerously close to the propeller. If your engine's cooling system is thermostatically controlled it may take a few moments for the thermostat to open, allowing a flow of water from the engine.

SPARK PLUGS

Spark plugs need to be cleaned and re-gapped periodically. Spark plugs are usually cleaned after 100 hours of operation, but some manufacturers recommend cleaning after as little as 50 hours. Since spark plugs operate under severe conditions, they become fouled easily. Knowing the signs of fouling can often tell about the engine's overall condition. Common problems that develop with spark plugs are carbon fouling, oil fouling, gas fouling, lead fouling, burned electrodes, chipped insulator, and splash fouling.

Normal spark plugs have light tan or gray deposits but show no more than .005″ increase in the original spark gap. These can be cleaned and reinstalled in the engine.

Worn out plugs have tan or gray colored deposits and show electrode wear of .008″ to .010″ greater than the original gap. Throw such plugs away and install new ones.

Oil fouling is indicated by wet, oily deposits. This condition is caused when oil is pumped (by the piston rings) to the combustion chamber. A hotter spark plug may help but the condition may have to be remedied by engine overhaul.

Gas fouling or fuel fouling is indicated by a sooty, black deposit on the insulator tips,

the electrodes and shell surfaces. The cause may be excessively rich fuel mixture, light loads, or long periods at idle speed.

Carbon fouling is indicated if the plug has dry, fluffy, black deposits. This condition may be caused by excessively rich fuel mixture; improper carburetor adjustment, choke partly closed, clogged air cleaner. Also, slow speeds, light loads, long periods of idling and the resulting "cool" operating temperature may cause deposits not to be burned away. A hotter spark plug may correct carbon fouling.

Lead fouling is indicated by a soft, tan, powdery deposit on the plug. These deposits of lead salts build up during low speeds and light loads. They cause no problem at low speed but at high speeds, when the plug heats up, the fouling will often cause the plug to misfire, thus limiting the engine's top performance.

Burned electrodes are indicated by thinned out, worn away electrodes. This condition is caused by the spark plug overheating. This overheating can be caused by lean fuel mixture, low octane fuel, cooling system failure, or long periods of high speed heavy load. A colder plug may correct this trouble.

Splash fouling may occur if accumulated cylinder deposits are thrown against the spark plugs. This material can be cleaned from the plug and the plug reinstalled. If spark gap tools and or pliers are not properly used, the electrode can be misshaped into a curve.

CLEANING SPARK PLUGS

If inspection indicates that the spark plug needs cleaning it can be cleaned within a few minutes. First, wire brush the shell and threads, not the insulator and electrodes. Then wipe the plug with a rag that has been saturated with solvent. This will remove any oil film on the plug. Next, file the sparking surfaces of the electrodes with a point file. Finally, regap the electrodes, setting according to specifications. Check the gap with a wire spark gap gage. Most service stations have an abrasive blast machine that is good for cleaning around the electrodes and insulator. Do not allow any foreign material to fall into the cylinder while the spark plug is out. And

NORMAL SPARK PLUG

WORN OUT SPARK PLUG

CARBON--FOULED SPARK PLUG

CHIPPED INSULATOR ON SPARK PLUG

OVERHEATED SPARK PLUG

OIL–FOULED SPARK PLUG

SPLASH-FOULED SPARK PLUG

BENT SIDE ELECTRODE

Fig. 8-2 Normal and Damaged Spark Plugs.

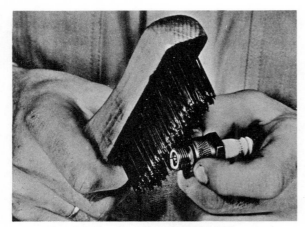

Fig. 8-3 Wire Brush the Shell and Threads.

Fig. 8-5A Regap the Plug.

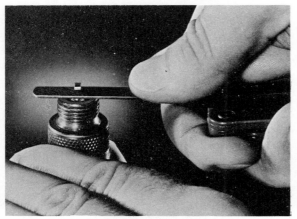

Fig. 8-4 File Electrodes.

Fig. 8-5B Check the Spark Gap.

also, do not forget the spark plug gasket when you reinstall the spark plug. This copper ring provides a perfect seal.

Of course, cleaning cannot repair a broken, cracked, or otherwise severely damaged spark plug. Cleaning is designed to maintain good operation and prolong spark plug life. It is not uncommon for small engine manufacturers to recommend the installation of a new spark plug at the beginning of each season. On a single-cylinder engine, the spark plug is relatively more critical than on a six- or eight-cylinder engine.

AIR CLEANER

The air cleaner serves the important function of cleaning the air before it is drawn through the carburetor and into the engine. Small abrasive particles of dust and dirt are trapped in the air cleaner. The air cleaner should be periodically cleaned: just how often depends mainly on the atmosphere in which the engine is operating. In a dusty atmosphere, a garden tractor used in a dry garden for example, the air cleaner should be cleaned every few hours. Under normal conditions, the air cleaner should be cleaned every twenty-five hours or sooner.

Fig. 8-6 About This Much Abrasive Material Would Enter a Six-cylinder Engine Every Hour if the Air Cleaner Were Not Used.

Air cleaners can be classified as (A) oil-type, which includes (1) oil bath air cleaner, and (2) oil-wetted polyurethane (foam) filter element, and (B) dry-type, which includes (1) foil, moss or hair element, (2) felt or fibre hollow element, and (3) metal cartridge air cleaner.

OIL BATH AIR CLEANER

The oil bath air cleaner carries a small amount of oil in its bowl. The level of this oil should be checked before each use of the engine. If the oil level is low, fill it <u>to the mark</u>, not above. Use the same type of oil as is used in the crankcase. As this cleaner works, both the oil and filter element will become dirty. To clean an oil bath air cleaner, disassemble the unit and then pour out the dirty oil. Wash the bowl, cover, and filter element in solvent; dry all parts; refill the bowl to the correct level and then reassemble the unit.

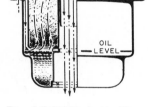

Fig. 8-7 Oil Bath Air Cleaner.

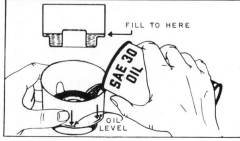

Fig. 8-8 Clean, Fill, Reassemble Air Cleaner (Foam).

OIL-WETTED POLYURETHANE FILTER ELEMENT AIR CLEANER

This type of air cleaner is perhaps the most common. Clean it by washing the element in kerosene, liquid detergent and water, or approved solvent. Dry the element by squeezing it in a cloth or towel. Apply considerable oil to the foam and work it throughout the element. Squeeze out any excess oil and then reassemble the air cleaner.

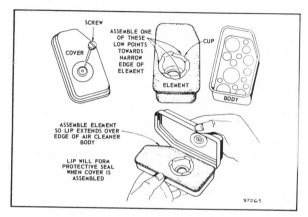

Fig. 8-9 Oil Foam Air Cleaner.

FOIL, MOSS, OR HAIR ELEMENT AIR CLEANERS

This type of element is also washed in solvent. After washing, allow the element to dry, then return it to the air cleaner body.

FELT OR FIBRE CYLINDER ELEMENT AIR CLEANER

These air cleaners should be cleaned by blowing compressed air through them in the opposite direction from normal flow. Dirt and dust particles will be blown out of the element.

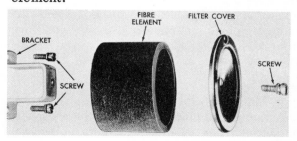

Fig. 8-10 Fibre Element Type Air Cleaner.

METAL CARTRIDGE AIR CLEANER

These air cleaners are cleaned by tapping or shaking to dislodge dirt accumulations.

Air cleaner elements are replaceable parts. If they wear out or become excessively clogged, they should be thrown away and replaced with a new filter element. As a simple test to determine if the air cleaner is badly clogged: (1) run the engine with the air cleaner removed; (2) with engine still running, replace the air cleaner; (3) notice if the engine speed remained constant or dropped down. Any noticeable drop in speed would probably indicate that the filter element is clogged.

Periodically inspect the engine for loose parts and do not neglect lubricating the engine's associated linkages, gear reduction units, chains, shafts, wheels, or the machinery the engine powers. These requirements vary greatly from engine to engine. On an outboard engine remember that the lower unit requires special lubrication.

Fig. 8-11 Metal Cartridge
Type Air Cleaner.

WINTER STORAGE AND CARE

Putting up an engine at the end of the season should be done carefully. Here are several items that should be accomplished:

1. Drain the fuel system.
2. Inject oil into the cylinder (upper cylinder).
3. Drain the crankcase (four cycle).
4. Clean the engine.
5. Wrap the engine in a canvas or blanket and store in a dry place.

The entire fuel system should be drained; tank, fuel lines, sediment bowl, and carburetor. Gasoline that is stored over a long period of time has a tendency to form gummy deposits. These deposits could impare and block the fuel system if gasoline stands in the system for a long time. Do not plan to hold over gasoline from season to season. Purchase a new supply when you return the engine to use.

Many manufacturers recommend that a small amount of oil be poured into the cylinder prior to storage. Remove the spark plug and pour in the oil (about a tablespoon in most cases), then turn the engine over by hand a few times to spread the oil evenly over the cylinder walls. Finally, replace the spark plug.

Some manufacturers suggest that the oil be introduced into the cylinder while the engine is running at a slow speed and just before the engine is stopped. To do this, the air cleaner is removed and a small amount of oil is poured in. The oil goes through the carburetor and on into the combustion chamber. When dark blue exhaust is produced the engine can be stopped.

The crankcase should be drained while the engine is warm so the oil will drain more readily. In some cases the crankcase is then refilled but on most engines it need not be refilled until the engine is returned to service.

Any unpainted surfaces that might rust should be lightly coated with oil before storage. Linkages should be oiled. The radiating fins and the engine in general should be cleaned. Do not put too much oil on the engine as it will collect dirt and be difficult to remove.

Finally, wrap the engine in a canvas or dry blanket and store the engine in a dry place. A garage, dry shed, or dry basement is good. Do not allow the engine to be left outside, exposed to the elements.

RETURNING THE ENGINE TO SERVICE

Upon returning the engine to service, refill the gas tank with new gasoline. Before filling the gas tank check to see if any water has condensed in the tank. If so, drain the water completely.

Check for condensation in the crankcase and refill the crankcase with new oil of the correct type. Also, clean the air cleaner and refill the bowl with oil if it is an oil bath type.

Check the spark plug; clean and regap it. Many manufacturers of small engines recommend that a new plug be installed at the beginning of each season to insure peak performance throughout the season.

CHECK OUT - BEFORE STARTING ENGINE

1. Gas tank full.

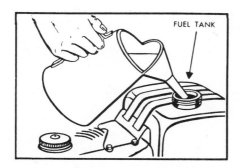

Fig. 8-12 Fill Gas Tank.

2. Oil in crankcase at correct level.

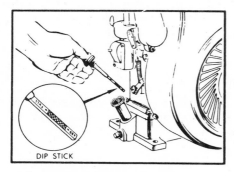

Fig. 8-13 Check Oil in Crankcase.

3. Fuel shutoff valve open.

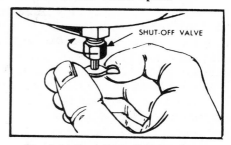

Fig. 8-14 Fuel Shut-off Valve Open.

4. Choke closed.

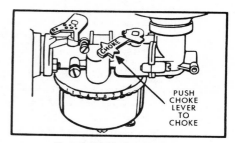

Fig. 8-15 Choke Closed.

5. Spark plug shorting bar off spark plug.

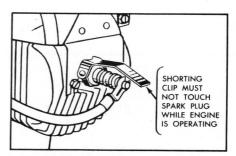

Fig. 8-16 Spark Plug Shorting Bar Off Spark Plug.

6. Air cleaner clean.

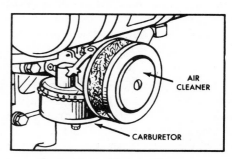

Fig. 8-17 Air Cleaner Clean.

SAFETY WITH GASOLINE ENGINES

Gasoline engines are made to be our friends; in work and in play. Through carelessness and disregard for simple safety rules a gasoline engine can be a dangerous killer. A list of persons injured or maimed through carelessness is a long one. Follow the safety rules listed below; a few minutes of time saved on a dangerous shortcut is never worth the price of an accident.

SAFETY RULES

1. Do not operate a gasoline engine in a closed building. The exhaust of an engine contains carbon monoxide, a deadly poison that is colorless and odorless.

2. Do not fill gas tank when the engine is running or hot. If gasoline is spilled on a hot engine, a fire or explosion can result. If you have not accomplished your job and the engine needs more gas, stop the engine and let it and yourself cool off before refilling the tank. It is a good idea to start each job with a full gas tank.

3. Do not make any adjustments to machinery being driven by an engine without first stopping the engine and removing the high-tension lead from the spark plug. It is possible to accidentally turn the engine over and start it while working on machinery, especially if the engine is warm. With the spark plug cable removed the engine cannot start.

4. Keep gasoline in a red gasoline can, in the garage or a tool shed, away from fire or open flame. Gasoline fumes are explosive; if you are working around open gasoline, be sure that no one is smoking.

Any piece of machinery demands the use of common sense precautions. In the case of the rotary power mower and other cutting machines, keep children out of the way. Do not point the grass discharge chute toward anyone. It is possible for the blade to throw a stick, stone, or other object with great force. Naturally, keep your hands and feet away from a rotating blade.

SPECIAL CONSIDERATIONS FOR OUTBOARD MOTORS

WINTER STORAGE

Generally, winter storage for outboard motors is similar to that for other engines. One notable difference is the cooling system. Being water cooled, all water must be removed from the cooling system. If a water filled cooling system on an idle engine is exposed to freezing weather, cracked water jackets and other major damage can result.

To remove the water from the system, remove the engine from the water, set the speed control on "stop" (to prevent accidental starting), and turn the engine over several times by hand. This will allow the water to drain from the water jackets and passages.

SALT WATER OPERATION

Although most engines are treated with anti-corrosives, not all engines are corrosive proof. If an engine normally used in salt water is not to be used for a while, its cooling system should be flushed with fresh water to prevent any possible corrosion.

USE IN FREEZING TEMPERATURES

If the engine is being used during freezing temperatures, care must be taken not to allow the cooling water to freeze in the engine or lower unit during an idle period. Of course while the engine is operating there is no danger of freezing.

GENERAL STUDY QUESTIONS

1. Explain what routine care and maintenance is and why it is important.
2. Where can routine care and maintenance information be found?
3. What are the four basic points to be considered in routine care and maintenance?
4. How often should the oil in the crankcase be checked?
5. How often should the oil in the crankcase be changed (Briggs-Stratton for example)?
6. What are the two types of cooling systems?
7. How often should the air-cooling system be cleaned?
8. How is an air-cooling system cleaned?
9. How do you care for a water-cooling system?
10. How often should spark plugs be cleaned?
11. What is the purpose of cleaning a spark plug?
12. List several types of spark plug fouling.
13. What are the steps in cleaning a spark plug?
14. What is the purpose of the air cleaner?
15. How often should air cleaners be cleaned?
16. What are the two basic types of air cleaners?
17. Discuss the cleaning of the several types of air cleaners.
18. List the general steps in winter storage of engines.
19. List the general steps in returning the engine to service.
20. List four safety rules for gasoline engines.
21. What special considerations must be made for outboard motors?

CLASS DISCUSSION TOPICS

- Discuss the importance of routine care and maintenance.
- Discuss the importance of care in winter storage.
- Discuss the types of cooling systems and their maintenance.
- Discuss the types of spark plug fouling.
- Discuss the importance and function of the air cleaner.
- Discuss the various types of air cleaners and how each might be used.
- Discuss the importance of safety with gasoline and gasoline engines
- Discuss your state laws in regard to gasoline storage.

CLASS DEMONSTRATION TOPICS

- Demonstrate cleaning and care of cooling systems.
- Demonstrate cleaning spark plugs.
- Demonstrate cleaning air cleaners.
- Demonstrate putting an engine into winter storage.
- Demonstrate returning an engine to service after winter storage.
- Demonstrate the check-out of an engine prior to starting.
- Demonstrate routine care and maintenance on an outboard engine.
- Demonstrate routine care and maintenance on a four-stroke cycle engine.

Unit 9

TROUBLESHOOTING - TUNE-UP - RECONDITIONING

Troubleshooting is the intelligent, step-by-step process of locating engine trouble. The troubleshooter examines and/or tests the engine to determine the cause of its "sickness". To engage in this process a person must thoroughly understand the "how" and "why" of engine operation. Really, troubleshooting is more of a mental activity than a physical one.

The troubleshooter first establishes the engine's symptoms. Then he checks the most common causes for these symptoms. If the solution is not found, he moves on and explores another possible cause for the engine's problem. The possible causes for engine failure must quickly be narrowed down, because there are hundreds of afflictions that can creep into an ailing engine.

An experienced mechanic will do his troubleshooting quickly and with little apparent effort. It would seem as if some sort of intuition guides him to the solution of the problem. Actually, it is his years of experience and his thorough understanding of engine operation that allows him to work with such competence.

The engine owner can do his own troubleshooting when he has an understanding of engine operating principles. Many "Owners Manuals" contain a troubleshooting chart that is a great assistance. With the aid of an engine troubleshooting chart, the engine owner or operator can locate and correct many engine difficulties without calling in a professional mechanic.

Even in the event your efforts at troubleshooting should fail and you find it necessary to consult an expert, you will derive a measure of satisfaction in being able to intelligently discuss your engine's troubles.

Troubleshooting is not a mysterious subject, but it does take a knowledge of engine operation and you must use common sense. For example, if the engine does not start there are two very logical reasons: either (1) fuel is not getting into the combustion chamber, or (2) fuel is being provided but it is not being ignited. Therefore, we look for the trouble in either the fuel system or the ignition system. Lubrication or cooling systems would have little bearing on the ability of the engine to start, so these systems are not inspected.

The troubleshooter starts with simple and most frequent causes of trouble. In inspecting the fuel system, the first step is to check for fuel in the tank. If this is not the cause of the trouble, then check the fuel shutoff valve, continuing logical steps until the trouble is found. It would be a mistake to completely disassemble the carburetor or magneto as the first step in troubleshooting.

Look at the Troubleshooting Charts in the Appendix, pages 270-275, and discover how many of the common engine troubles can be corrected by an owner who has the normal household tools at his disposal.

ENGINE TUNE-UP

Engine tune-up does not involve major engine repair work. Rather, it is a process of cleaning and adjusting the engine so that it will give top performance. Tune-up can be done by an experienced engine owner or it can be done by a mechanic.

A typical tune-up procedure might include the following steps:

1. Inspect air cleaner; clean and reassemble air cleaner. Is the cleaner damaged in any way? Is the filter element too clogged for cleaning? If so, replace the cleaner and/or element. Clean the unit according to the manufacturer's instructions.

2. Clean the gas tank, fuel lines, and any fuel filters or screens.

3. <u>Check compression.</u> This can be done by slowly turning the engine over by hand; as the piston nears top dead center, considerable resistance should be felt. As top dead center is passed the piston should "snap" back down the cylinder. A more thorough test can be made with a compression gage. The manufacturer's repair manual should be consulted for normal and minimum acceptable compression pressure.

Fig. 9-1 Checking Compression with a Compression Gage.

4. <u>Check spark plug</u>; clean, regap or replace. Remove high-tension lead from plug and hold it about 1/8" from the plug base. Be sure to hold it on the insulation. Turn the engine over. If a good spark jumps to the plug, magneto is providing sufficient spark. Now replace the high-tension lead and lay the plug on a bare spot on the

engine. Turn engine over. If a good spark jumps at the plug electrodes, the plug is good. This is not an absolute test but is a good indication. If the plug is questionable, replace it.

5. <u>Check operation of the governor.</u> Be certain governor linkages do not bind at any point.

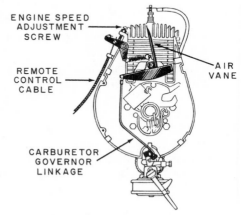

Fig. 9-2 Air Vane Governor Parts.

6. <u>Check magneto.</u> On most engines, the flywheel must be removed for access to the magneto parts. Use a flywheel or gear puller, if available. Otherwise, pop the flywheel loose by removing the flywheel nut and delivering a sharp hammer blow to a lead block held against the end of the crankshaft.

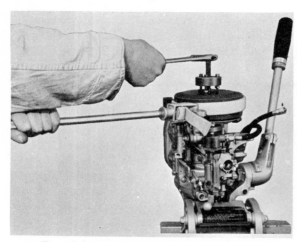

Fig. 9-3 Removing a Flywheel Using a Special Puller.

Fig. 9-4 Removing the Flywheel with a Soft Hammer.

Another method is to back off the flywheel nut until it is about 1/3 off the crankshaft, then strike it sharply. In no case should you hammer on the end of the crankshaft as this may damage the threads.

Adjust the breaker point gap and check the condenser and breaker point terminals for tightness. Breaker points will be set for .020″, in most cases, although the correct setting may vary according to the manufacturer. Turn the engine over until the breaker points reach their maximum opening, then check with a flat feeler gage. Adjust the points if the setting is incorrect. Points can be cleaned if necessary. If the points are pitted or do not line up, they should be replaced.

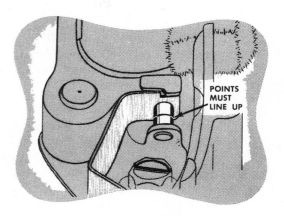

Fig. 9-6 Breaker Points Must Line Up.

7. Fill crankcase with clean oil of the correct type. (four-stroke cycle engines)

8. Fill gasoline tank with regular gasoline. (four-stroke cycle engines)

   Fill gasoline tank with correct mixture of gasoline and oil. (two-stroke cycle engines)

9. Start engine.

10. Adjust the carburetor for peak performance.

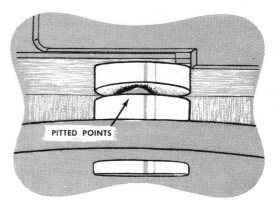

Fig. 9-7 Replace Pitted Breaker Points.

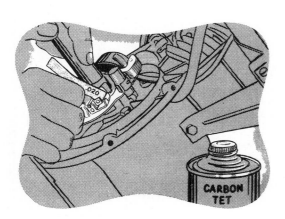

Fig. 9-5 Clean and Check
Breaker Points.

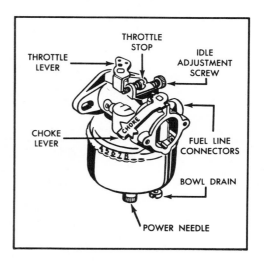

Fig. 9-8 Carburetor Parts.

RECONDITIONING

Reconditioning or overhauling an engine is generally the job for the mechanic; one who has the tools and know-how to do the job correctly. However, much reconditioning and overhauling can be done by the amateur who has the essential tools and the mechanic's workmanlike approach. Care and precision are vitally important; slipshod workmanship is never tolerated in engine work. Every part must be in place correctly, none missing or left over.

These points are stressed because at times the novice mechanic with an engine is like the small child taking a watch apart to see just what makes it tick. Rarely does the child get the watch back together so it works. Engine parts are larger but there are hundreds of parts that make up any engine and they all must go back together correctly.

Before you begin disassembly of an engine, provide a spot to receive the parts as they are removed. One good method is to lay out a large sheet of paper and as the parts are placed on the paper, label them so they will not be lost. It is often difficult to remember just where a part came from when forty or fifty pieces are laid out, and it may be several days before you reassemble the engine.

A "Mechanic's Handbook" or "Service Manual" is essential for top quality overhaul work. These books can sometimes be obtained direct from the engine manufacturer or borrowed from a mechanic. In only a few instances is this type of book supplied with the new engine. The service manual goes into great detail, fully explaining each step of overhaul or reconditioning. Allowances, clearances, torque data, and other specifications are given as well. Without the use of the service manual there is too much guesswork left to the amateur mechanic. The amateur mechanic needs the service manual for an engine before repair work is started.

CLINTON ENGINES TORQUE DATA — INCH POUNDS
"Red Horse"

		1600 A1600 A1690	1800 1890	2500 A2500	B2500 B2590 2790
Bearing Plate P.T.O.	Min.	160	160	160	160
	Max.	180	180	180	180
Back Plate to Block	Min.	70	70	70	70
	Max.	80	80	80	80

CLINTON ENGINES SERVICE CLEARANCES
"Red Horse"

		1600	A1600	A1690	1800	1890	2500	B2500	B2590	A2500	2790
Piston Skirt Clearance	Min.	.007	.007	.007	.007	.007	.0065	.0065	.0065	.005	.005
	Max.	.009	.009	.009	.009	.009	.0085	.0085	.0085	.007	.007
	Rework	.010	.010	.010	.010	.010	.010	.010	.010	.0085	.0085
Ring End Gap	Min.	.007	.007	.007	.007	.007	.010	.010	.010	.010	.010
	Max.	.017	.017	.017	.017	.017	.020	.020	.020	.020	.020
	Rework	.025	.025	.025	.025	.025	.028	.028	.028	.028	.028

CLINTON ENGINES TOLERANCES AND SPECIFICATIONS
"Red Horse"

	1600	A1600	A1690	1800	1890	2500	B2500	B2590	A2500	2790
Cylinder - Bore	2.8125	2.8125	2.8125	2.9995	2.9995	3.1245	3.1245	3.1245	3.1245	3.1245
	2.8135	2.8135	2.8135	3.0005	3.0005	3.1255	3.1255	3.1255	3.1255	3.1255
Skirt - Diameter	2.8045	2.8045	2.8045	2.9915	2.9915	3.117	3.117	3.117	3.1185	3.1185
	2.8055	2.8055	2.8055	2.9925	2.9925	3.118	3.118	3.118	3.1195	3.1195

Fig. 9-9 Torque Data, Clearances, Tolerances, and Specifications as Excerpted from a Service Manual.

Unless the engine owner is a skilled mechanic and has an adequately equipped repair shop, he will not be able to perform all types of engine repair and overhaul. Most owners would not find it practical to purchase the equipment necessary to do all types of overhaul and repair. Professional repairmen may have anywhere from $500 to several thousand dollars invested in tools and equipment: magneto testers, valve seat resurfacers, air compressors, steam cleaners, special sharpeners, grinders, lapping stands, special factory tools, general repair tools, repair parts, etc.

However, a modest investment in tools will enable a person to perform many repair jobs. Most home workshops have a number of the basic tools that are necessary and with the addition of some specialized tools he will have a reasonably good set of tools. A tentative list of necessary tools includes:

1. Screwdriver (various sizes)
2. Combination wrenches (set)
3. Adjustable wrench
4. Socket set
5. Torque wrench
6. Deep well spark plug socket
7. Needle nose pliers
8. Feeler gage
9. Spark gap gage
10. Piston ring compressor
11. Piston ring expander
12. Valve spring compressor

Additional tools:

1. Flywheel puller
2. Compression gage
3. Tachometer
4. Valve grinder - hand operated
5. Arbor press

Reconditioning or overhaul of an engine involves four basic steps: (1) disassembly, (2) inspection of parts, (3) repair or replacement of worn or broken parts, and (4) reassembly. The following disassembly procedure is a general guide for a four-stroke cycle engine.

1. Disconnect spark plug lead - remove spark plug.
2. Drain fuel system - tank, lines, carburetor.
3. Drain oil from crankcase.
4. Remove air cleaner.
5. Remove carburetor.
6. Remove metal air shrouding, gas tank and recoil starter.
7. Remove flywheel.
8. Remove breaker assembly and push rod.
9. Remove magneto plate assembly.
10. Remove breather plate assembly (valve spring cover).
11. Remove cylinder head.
12. Remove valves.
13. Remove base.
14. Remove piston assembly.
15. Remove crankshaft.
16. Remove camshaft and tappets.
17. Remove mechanical governor.

In the following discussion, disassembly, inspection, repair, and reassembly will be discussed for each engine part. The material is, of course, general, and the illustrations cover several different makes of engines. In actual practice a service manual would be followed for these steps. However, this information is an excellent source for the general procedure followed by most manufacturers.

(1) disconnect spark plug lead - remove spark plug, (2) drain fuel system - tank, lines, carburetor, (3) drain oil from crankcase, (4) remove air cleaner, (5) remove carburetor, and (6) remove metal air shrouding. Gas tank and recoil starter are steps that have been thoroughly covered in earlier units and not of sufficient difficulty to be discussed at greater length. Be certain to inspect any parts removed for damage and wear. Replacement or repair may be indicated.

Removal of the flywheel is necessary to gain access to the ignition system parts. If a flywheel puller is available, it is best; however, the flywheel can be removed by striking the end of the crankshaft with a plastic or soft hammer. Use care not to damage the end of the crankshaft. The key on the crankshaft should not be lost. Keep it on the crankshaft or in a safe place. When reassembling the flywheel onto the crankshaft, carefully fit the key into the keyway.

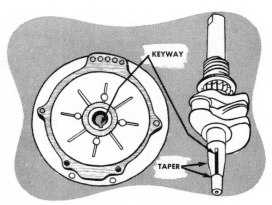

Fig. 9-10 Keyway on Tapered End of the Crankshaft and the Flywheel.

MAGNETO PARTS

The magneto parts - high-tension coil, breaker points, condenser - may need to be removed for checking or replacement. The coil may be visually inspected for cracks and gouges in insulation, evidence of overheating, and the condition of the leads where they go into the coil. If an ignition coil tester is available, the coil can be checked for (1) firing check, (2) leakage check, (3) secondary continuity check, and (4) primary continuity check. This checking should be done using the procedure recommended by the manufacturers.

Breaker points can be visually inspected for pitting, alignment, and contact surface.

The condenser can be visually inspected for dents, terminal lead damage, and broken mounting clip. A condenser tester is used to check the condenser for capacity, leakage, and series resistance. Follow the test procedure suggested by the test equipment manufacturer.

Fig. 9-11 Checking a Condenser.

When the magneto is reassembled, the laminated iron core must be close to the flywheel magnet but not touching it. Generally, the closer the better, but not close enough to rub. This air gap is specified by the manufacturer. Too large an air gap can cause faulty magneto operation.

ENGINE TIMING

Engine timing refers to the magneto timing to the piston: the position of the piston just as the breaker points start to open. Timing is set at the factory but it is possible for timing to cause engine trouble. If the breaker points open too late in the cycle, power is lost; if the breaker points open too early in the cycle, detonation may result. Improper engine timing could be caused by the crankshaft and camshaft gears being installed one tooth off, spark advance mechanism stuck, breaker points incorrectly set, magneto assembly plate loose or slipped, or rotor incorrectly positioned. These causes of trouble depend on the engine and type of magneto.

The position of the piston for timing a magneto varies from engine to engine. The piston may be at top dead center (TDC) or slightly before top dead center (BTDC). Check engine specifications for this information.

One common method of checking timing is to locate TDC. This is the point where the piston does not seem to move as the flywheel is rotated. A dial indicator can be used for greater accuracy. With TDC located, place a reference mark on the flywheel and the magneto assembly plate. Now remove the flywheel. Rotate the crankshaft until the breaker points just open (.001″). Carefully replace the flywheel and put another mark on the magneto plate, aligned with the flywheel reference mark. The difference between the marks on the magneto plate is the magneto timing to the piston. This can be figured in degrees by dividing the number of flywheel vanes into 360°. Twenty vanes; each vane 18°. The correct number of degrees of firing before top dead center is found in the engine specifications.

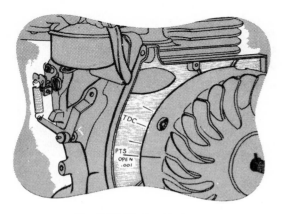

Fig. 9-12 Timing Marks.

If the engine has a magnetic rotor magneto, the rotor and armature must be timed to the piston. Timing is correctly set when the engine leaves the factory, but if the armature has been removed or the crankshaft or cam gear replaced, it is necessary to retime the rotor. Breaker points are first set correctly (.020″). Rotor is on the shaft correctly and tightened. Armature is mounted but mounting screws are not tight. Turn the crankshaft until the breaker points just start to open (place a piece of tissue paper between the points to detect when the points "let go"). Now turn the armature slightly until the timing marks on the rotor and the armature line up.

Two-cycle engine timing is quite similar. One common procedure is to remove the spark plug and, using a ruler and straightedge across the head, locate top dead center. With TDC located, back the piston down the cylinder the

correct distance (check manufacturer's specifications for the measurement before TDC). At this point, the breaker points should just begin to open. If timing is incorrect, loosen the stator plate setscrew and rotate the stator plate slightly until the points just begin to open. Retighten the stator plate setscrew.

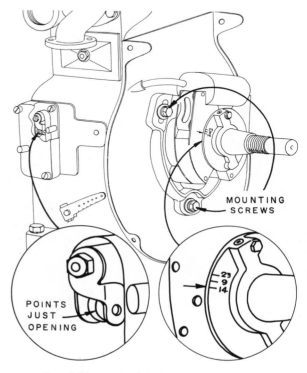

Fig. 9-13 Timing the Rotor and Armature to the Piston. The Numbers Represent Model Numbers.

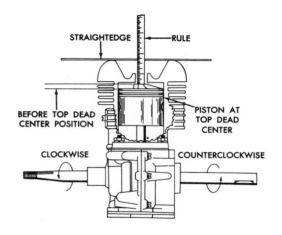

Fig. 9-14 Two-cycle Engine Timing.

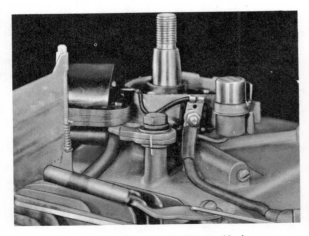

Fig. 9-15 Two-cycle Timing Marks.

Fig. 9-17 Tightening the Cylinder Head Bolts with a Torque Wrench.

Fig. 9-16 Two-cycle Timing Marks.

CYLINDER HEAD

The cylinder head must be removed if work is to be done on the piston, piston rings, connecting rod, valves, or if the combustion chamber is to be cleaned. The head screws or nuts should be removed and set aside. After the head is removed, the head gasket should be removed and discarded. Before reassembly, carbon deposits should be scraped off and the head cleaned. A new head gasket should be installed and the cylinder head screws or nuts should be tightened in the correct sequence. To be certain of the correct degree of tightness a torque wrench should be used (generally 14 to 18 ft./lbs).

VALVES

The valves can easily be inspected. Inspection may show that they are stuck, burned, cracked, or fouled with carbon. Also, valve stems and valve guides may be worn or the valves and valve seats may need to be reground. Further, tappet clearance may be wrong. The average engine owner may be able to do some valve work himself. However, extensive work or a complete renewal of the valve system might best be left to a mechanic with the tools and experience necessary.

A complete valve job might include:

1. Installing new valves

2. Installing new valve guides

3. Installing new valve seats

4. Installing new valve springs

5. Grinding valve seats

6. Lapping valves

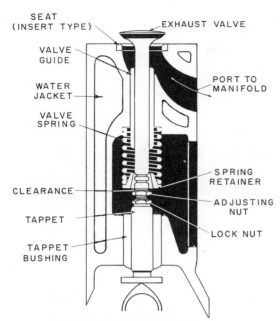

Fig. 9-18 Complete Valve Train.

Fig. 9-19 Checking Tappet Clearance.

One of the first check points is the tappet clearance (space between tappet and end of valve). This clearance is checked with a feeler gage. On most small engines the clearance can be enlarged by grinding a small amount from the end of the valve. If the clearance is already too large the valve would have to be replaced. Some small engines do have adjustable tappet clearances.

To remove the valves, first compress the valve spring, then flip off or slip out the valve spring retainers, sometimes called keepers. Pull the valve out of the engine. Valve spring and associated parts will also come out.

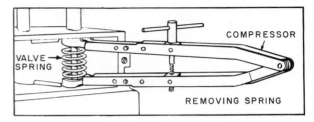

Fig. 9-20 Using a Valve Spring Compressor to Remove the Valve Spring.

With the valve out, the valve stem, face, and head can be closely inspected and cleaned. The valve guide and valve seat can also be inspected.

Many engines have replaceable valve guides. If these are worn or otherwise damaged, they must be pressed out using an arbor press or carefully driven out with a special punch. Also, the exhaust valve seat is removable on many engines; if the seat is beyond regrinding, remove it and reinstall a new valve seat. A special valve seat extracting tool is used for this job.

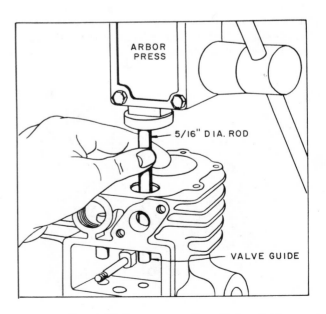

Fig. 9-21 Removing a Worn Valve Guide.

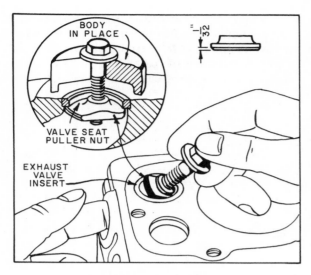

Fig. 9-22 Removing the Exhaust Valve
Seat with a Special Puller.

Fig. 9-24 Power Drill Operated Valve Seat
Grinder.

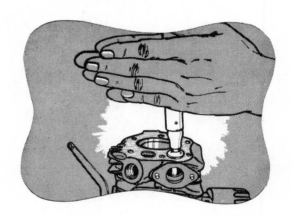

If the valves and/or valve seats need re-grinding, this can be done with special valve grinding equipment. However, in some cases hand valve grinders can be used.

If either or both of these parts have been replaced or reground, the parts must be lapped to provide the perfect seal necessary for valve operation. When lapping, use a small amount of lapping compound. Rotate the valve against the seat a few times until the compound produces a dull finish on the valve face. Do not lap the valves too heavily.

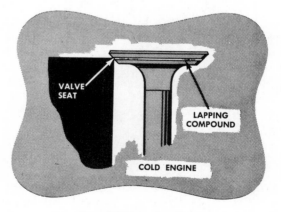

Fig. 9-25 Lapping Valves.

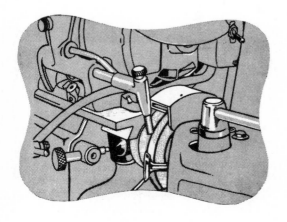

Fig. 9-23 Valve Grinder.

When the valves are replaced, oil the stems and be certain that the exhaust valve goes in the exhaust side, intake valve in intake side.

Remove the engine from the base or sump. This is done by loosening the bolts and breaking the seal. In most cases, a new gasket will have to be installed upon reassembly.

Fig. 9-26 Engine Base, Gasket, Engine.

To remove piston assembly, remove the connecting rod cap and push the piston assembly up out of the engine. Mark the piston so it can be reinstalled the same way it comes out. The cylinder should be checked for score marks. Scoring in the area of ring travel will cause excessive oil consumption and reduced engine power. Also the cylinder size should be checked with a cylinder gage. If the cylinder appears to be in good condition, it can be deglazed with a finish hone to prepare cylinder for new rings.

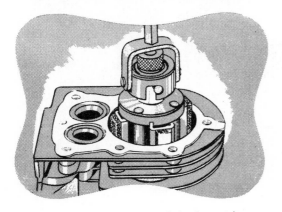

Fig. 9-27 Deglazing a Cylinder with a Finish Hone.

If the piston size and condition has checked out all right, the piston rings should be checked next. Check the edge gap with a feeler gage, rings still on the piston. The correct edge gap clearance is found in the engine overhaul specifications. Remove the piston rings from the piston with a piston ring expander. Carefully put the ring in the cylinder and check the end gap with a feeler gage. Too little end gap may cause the ring to freeze when it becomes hot and expands. Too much end gap may allow "blow-by" and the resulting loss of power. Piston ring grooves should be cleaned to remove any carbon accumulations. New piston rings are normally installed during engine overhaul.

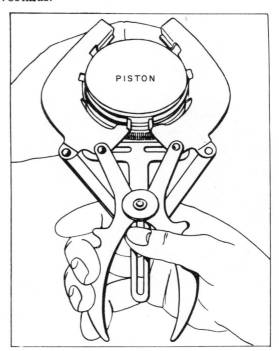

Fig. 9-28 Special Tool Used For Removing Piston Rings.

Fig. 9-29 Piston Ring Tool.

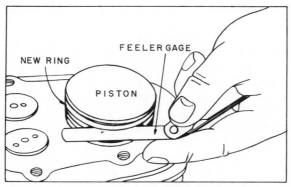

Fig. 9-30 Checking the Ring Groove with
a Feeler Gage.

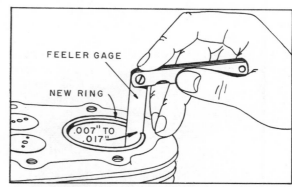

Fig. 9-31 Checking the Ring Gap with a
Feeler Gage. (End Gap)

Use a piston ring compressor to reinstall the piston assembly. Place the piston in the same way as it came out. Also, put the connecting rod cap on the same way as it came off - find the match marks. A torque wrench should be used to tighten the connecting rod cap.

The crankshaft should be removed and checked for scoring and any metallic pickup. The journal and crankpin should be checked with a micrometer for roundness. The gear and keyway should be checked for wear. In some cases the main ball bearings may come out with the crankshaft. Upon reinstallation, the bearings may have to be pressed into place; an arbor press is good for this job.

Fig. 9-32 Piston Ring Compressor.

Fig. 9-33 Torque Wrench Used to Correctly
Tighten Connecting Rod Cap.

Fig. 9-34 Pressing in the Main Ball Bearing.

Fig. 9-35 Reinstalling the Crankshaft.

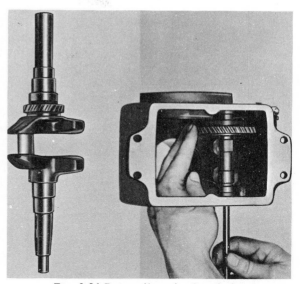

Fig. 9-36 Reinstalling the Camshaft.

The camshaft and valve tappets can also be removed for inspection and repair. The camshaft pin should be driven out with a drift punch from the power takeoff side of the engine. Upon reassembly of the camshaft and crankshaft, be certain to line up the timing marks.

GENERAL STUDY QUESTIONS

1. Define troubleshooting.
2. Of what value is a troubleshooting chart?
3. Can troubleshooting be done by the average engine owner or operator?
4. Define engine tune-up.
5. Define engine reconditioning or overhaul.
6. Can any reconditioning or overhaul be accomplished by the amateur mechanic?
7. Why is the careful layout of parts important during disassembly?
8. Why is a Mechanic's Handbook or Service Manual essential for the reconditioning of an engine?
9. Explain engine timing.
10. What might a complete valve job include?

CLASS DISCUSSION TOPICS

● Discuss the troubleshooting chart.
● Discuss the steps in engine tune-up.
● Discuss the importance of care, accuracy, etc. in engine reconditioning and overhaul work.
● Discuss the tools necessary for reconditioning work.
● Discuss the steps in engine reconditioning.

CLASS DEMONSTRATION TOPICS

▶ Demonstrate troubleshooting by putting troubles into an engine and having students observe engine operation and then troubleshoot the engine. Simple examples: no fuel in tank, poor compression, shorting bar on plug, fuel shutoff valve closed, spark plug lead loose, bad spark plug, etc.

▶ Demonstrate how to tune up an engine.

▶ Demonstrate the correct use of the basic tools.

▶ Demonstrate checking the magneto for damage or wear.

▶ Demonstrate timing an engine.

▶ Demonstrate the tightening sequence and correct torque on a cylinder head.

▶ Demonstrate removing and inspecting valves for wear and damage.

▶ Demonstrate checking tappet clearance.

▶ Demonstrate lapping valves.

▶ Demonstrate removing and inspecting piston assembly for damage and wear.

▶ Demonstrate removal and inspection of the camshaft and tappets.

HORSEPOWER - SPECIFICATIONS - BUYING CONSIDERATIONS

HORSEPOWER

Horsepower is the yardstick of the engine's power, its capacity to do work. James Watt, the inventor of the steam engine, devised the unit. He assumed that the average horse could raise 33,000 pounds one foot in one minute. An engine that can lift 33,000 pounds one foot in one minute is a one-horsepower engine. Of course, the engine can lift 8,250 pounds four feet in one minute, and still be delivering 1 hp.

As discussed in Unit 2, the formula for horsepower is:

$$hp. = \frac{Work}{Time\ (in\ minutes) \times 33,000}$$

Remember that work is the energy required to move a weight through a distance. For example: lifting one pound, one foot, is one foot-pound of work. Lifting 50 pounds, two feet, is 100 foot-pounds of work. The element of time is not a factor.

In horsepower the element of time is added. For example, a heavily laden boat with an outboard motor might cross a river in ten minutes. An identical boat with a larger motor might make the crossing in five minutes. The same amount of work has been accomplished by each engine but the larger horsepower engine did the work faster.

Horsepower terms are varied and often misleading. Persons speak of Rated hp., Developed hp., Brake hp., Indicated hp., Maximum hp., Continuous hp., Corrected hp., Frictional hp., Observed hp., etc. Most of these terms will be discussed in the following paragraphs.

Brake Horsepower is usually used by manufacturers to advertise their engine's power. Brake horsepower is measured either with a prony brake or with a dynamometer. The prony brake consists basically of a flywheel pulley, adjustable brake band, lever, and scale measuring device. With the engine operating, the brake band is tightened on the flywheel and the pressure or torque is transmitted to the scale. The readings on the scale and other data are used to calculate the horsepower.

The dynamometer is a more recent device for measuring horsepower. The electric dynamometer contains a dynamo and as the engine drives the dynamo the current output can be carefully recorded. The more powerful the engine, the more current is produced.

Frictional Horsepower is the power that is used to overcome friction in the engine. The parts themselves absorb a certain amount of power. The pistons account for the greatest friction loss.

Indicated Horsepower is the power that is actually produced by the burning gases within the engine. It does not take into account the power that is absorbed by or used to move the engine parts. It would be the sum of the brake horsepower plus the power used to drive the engine (frictional horsepower).

SAE or Taxable Horsepower is the horsepower used to compute the license fee for automobiles in some states.

$$SAE\ hp. = \frac{D^2 \times No.\ of\ Cylinder}{2.5}$$

D = Diameter of bore in inches

$$Example:\ \frac{(3.5 \times 3.5) \times 6}{2.5} = 29.4\ hp.$$

ESTIMATING HORSEPOWER

The following formula can be used to estimate the maximum horsepower of a four-stroke cycle engine.

$$hp. = \frac{D \times N \times S \times r.p.m.}{11,000}$$

D (Bore in Inches)
N (Number of Cylinders)
S (Stroke in Inches)
11,000 (Experimental Constant)

For a two-cycle engine, the formula can be used with 9000 as the experimental constant.

Engine Torque is a factor that relates to horsepower. Torque is the twisting force of the engine's crankshaft. Torque can be compared to the force a man uses to tighten a nut. At first a small amount of torque is used, but more and more torque is applied as the nut tightens; even when the nut stops turning, the man still may be applying torque. Motion is not necessary to have torque. Torque is measured in foot-pounds or inch-pounds. On an engine, the maximum torque is developed at speeds below the maximum engine speed. At top speeds, frictional horsepower is greater and volumetric efficiency is less.

Volumetric Efficiency relates to horsepower also. It refers to the engine's ability to breathe properly, that is, its ability to take in a full charge of fuel mixture in the short time allowed for intake. Engine design largely determines the volumetric efficiency of the engine. At high speeds the volumetric efficiency drops off. It can be increased with superchargers and turbochargers that force or blow air into the intake manifold. Also, volumetric efficiency can be increased by using multiple barrel carburetors (2 barrel and 4 barrel), the extra barrels opening up at high speed when the demand for air is greater.

Compression Ratio is another factor that affects horsepower. This is the relationship of the volume of the cylinder when the piston is at the bottom of its stroke compared to the volume of the cylinder (and combustion chamber) when the piston is at top dead center. For small gasoline engines it may be 6:1, for automobile engines, the compression ratio may be 8:1 or higher. The higher the compression ratio, the greater the horsepower delivered when the fuel mixture is ignited.

Piston Displacement is the volume of air the pistons displace from the bottom of their stroke to top dead center. Piston displacement also relates to horsepower. Generally, the greater the piston displacement the greater the horsepower. Most engines will deliver 1/2 to 7/8 horsepower per cubic inch displacement. High performance engines will develop about 1 horsepower per cubic inch displacement. Supercharged engines can deliver considerably more than 1 horsepower per cubic inch displacement.

Displacement equals Area of Bore × Stroke × Number of Cylinders, and is expressed in cubic inches.

SPECIFICATIONS

Below are specifications for several models of some of the leading manufacturers of engines. In considering the best engine for a certain job, engineers (and prospective owners too) study such information.

CUSHMAN MOTORS, Lincoln, Nebraska — Principal Usage: Utility Vehicles*

* Partial List of Models Model*	Rated hp.	R.p.m.	Bore	Stroke	Disp. cu. in.	Comp. Ratio	2 or 4 Cycle	Weight
100	9		3.5	2.25	21.58	6.85/1	4	
200	18		3.5	2.25	43.16	6.85/1	4	

KOHLER CO., Kohler, Wisconsin — Principal Usage: All engine powered equipment

* Other models too numerous to mention Model*	Rated hp.	R.p.m.	Bore	Stroke	Disp. Cu. In.	Comp. Ratio	2 or 4 Cycle	Weight
K91	4.1	4000	2 3/8	2	8.86	6.5/1	4	41
K161	7.0	3600	2 7/8	2 1/2	16.22	6.25/1	4	65
KV161	7.0	3600	2 7/8	2 1/2	16.22	6.25/1	4	65
L160	6.5	3600	2 7/8	2 1/2	16.22	6.25/1	4	106
K241	9.5	3600	3 1/4	2 7/8	23.9	6.00/1	4	105
K331	12.5	3200	3 5/8	3 1/4	33.6	6.25/1	4	173
K662	24.0	3200	3 5/8	3 1/4	67.2	6.00/1	4	250

JACOBSEN MANUFACTURING CO., Racine, Wisconsin
Principal Usage: Lawn Equipment

Model	Rated hp.	R.p.m.	Bore	Stroke	Disp. Cu. In.	Comp. Ratio	2 or 4 Cycle	Weight
J-125-H	2.25		2.0	1.5	4.71	5.5/1	2	
J-100-H	1.8		2.0	1.5	4.71	5.5/1	2	
J-125-V	2.25		2.0	1.5	4.71	5.5/1	2	
J-175-H	3.0		2.12	1.75	6.2	5.5/1	2	
J-175-V								
J-225-V	4.0		2.25	2.0	7.95	5.3/1	2	
J-321-V	3.0		2.12	1.75	6.2	5.0/1	2	
J-321-H								

BRIGGS AND STRATTON CORP., Milwaukee, Wisconsin 53201
Principal Usage: General Power Use

* Other models too numerous to mention

Model*	Rated hp.	R.p.m.	Bore	Stroke	Disp. Cu. In.	Comp. Ratio	2 or 4 Cycle	Weight
92500	3	3600	2 9/16	1 3/4	9.02		4	19.5
100900	4	3600	2 1/2	2 1/8	10.43		4	30.5
130900	5	3600	2 9/16	2 7/16	12.57		4	30.75
60100	2	3600	2 3/8	1 1/2	6.65		4	22.25
80100	2.5	3600	2 3/8	1 3/4	7.75		4	22.25
80300	3	3600	2 3/8	1 3/4	7.75		4	25.25
190400	8	3600	3	2 3/4	19.44		4	45

CLINTON ENGINES CORP., Maquoketa, Iowa
Principal Usage: General Power Use

* Other models too numerous to mention

Model*	Rated hp.	R.p.m.	Bore	Stroke	Disp. Cu. In.	Comp. Ratio	2 or 4 Cycle	Weight
A2100	2.25	3600	2 3/8	1 5/8	7.2		4	21 1/2
100	2.50	3600	2 3/8	1 5/8	7.2		4	23
4100	2.75	3600	2 3/8	1 7/8	8.3		4	21 1/2
3100	3.00	3600	2 3/8	1 7/8	8.3		4	23
V1000	3.25	3600	2 3/8	1 7/8	8.3		4	36
V1100	3.75	3600	2 3/8	2 1/8	9.5		4	36
B1290	4.00	3600	2 15/32	2 1/8	10.2		4	45
V1200	4.50	3600	2 15/32	2 1/8	10.2		4	40
A1600	6.30	3600	2 13/16	2 5/8	16.3		4	87
B2500	9.60	3600	3 1/8	3 1/4	25.0		4	103
2790	10.30	3600	3 1/8	3 1/4	25.0		4	103

LAWN BOY, Galesburg, Illinois 61401
Principal Usage: Lawn Mowers

* Other models too numerous to mention

Model*	Rated hp.	R.p.m.	Bore	Stroke	Disp. Cu. In.	Comp. Ratio	2 or 4 Cycle	Weight
C-10		4000	1 15/16	1 1/2	4.43	6.5/1	2	
C-12AA		4000	2 1/8	1 1/2	5.22	6.5/1	2	
C-18			2 3/8	1 1/2	6.65		2	
D-400			2 3/8	1 1/2	6.65		2	

WISCONSIN MOTOR CORP., Milwaukee, Wisconsin
Principal Usage: Heavy Duty Industrial

Model*	Rated hp.	R.p.m.	Bore	Stroke	Disp. Cu. In.	Comp. Ratio	2 or 4 Cycle	Weight
*Other models too numerous to mention								
ACN	6	3600	2 5/8	2 3/4	14.88		4	76
BKN	7	3600	2 7/8	2 3/4	17.8		4	76
AENL	9.2	3600	3	3 1/4	23		4	110
AEH	7.4	3200	3	3 1/4	23		4	130
AGND	12.5	3200	3 1/2	4	38.5		4	180
THD	18	3200	3 1/4	3 1/4	53.9		4	220
VE4	21.5	2400	3	3 1/4	91.9		4	295
VF4	25	2400	3 1/4	3 1/4	107.7		4	295
VH4	30	2800	3 1/4	3 1/4	107.7		4	310
VG4D	37	2400	3 1/2	4	154		4	410
VR4D	56.5	2200	4 1/4	4 1/2	255		4	775

GRAVELY TRACTORS, INC., Dunbar, West Virginia
Principal Usage: Gravely Tractors

Model	Rated hp.	R.p.m.	Bore	Stroke	Disp. Cu. In.	Comp. Ratio	2 or 4 Cycle	Weight
L	6.6	2600	3 1/4	3 1/2	29.0	5/1	4	296

O & R ENGINES, Los Angeles, California 90023
Principal Usage: General Power

Model	Rated hp.	R.p.m.	Bore	Stroke	Disp. Cu. In.	Comp. Ratio	2 or 4 Cycle	Weight
13B	1.0	6200	1.25	1.096	1.34	9/1	2	3.9
13A	1.0	6600	1.25	1.096	1.34	9/1	2	3.9
20A	1.6	7200	1.437	1.250	2		2	4.9

CHRYSLER OUTBOARD CORP. (Formerly West Bend) Hartford, Wisconsin
Principal Usage: Chain Saws, Scooters, Carts

Model	Rated hp.	R.p.m.	Bore	Stroke	Disp. Cu. In.	Comp. Ratio	2 or 4 Cycle	Weight
27824	3	4500	2	1 5/8	5.1	5.5/1	2	13 1/2
27825	3	4500	2	1 5/8	5.1	5.5/1	2	13 1/2
27852	3	4500	2	1 5/8	5.1	5.5/1	2	13 1/2
27854	3	4500	2	1 5/8	5.1	5.5/1	2	13 1/2
27612	5	5500	2 1/4	1 3/4	7.0	5.7/1	2	13 1/2
2760	5	5500	2 1/4	1 3/4	7.0	5.7/1	2	13 1/2

D. W. ONAN AND SONS, INC., Minneapolis, Minnesota
Principal Usage: Compressors, Truck Ref. Mowers, Go-Carts, Scooters, etc.
(Onan Generator units not listed)

Model	Rated hp.	R.p.m.	Bore	Stroke	Disp. Cu. In.	Comp. Ratio	2 or 4 Cycle	Weight
AS	5.5	3600	2 3/4	2 1/2	14.9	6.25/1	4	85
CCK	12.9	2700	3 1/4	3	50.0	5.50/1	4	148

TECUMSEH PRODUCTS CO., Grafton, Wisconsin
Principal Usage: General Power Use

Model*	Rated hp.	R.p.m.	Bore	Stroke	Disp. Cu. In.	Comp. Ratio	2 or 4 Cycle	Weight
* Other models too numerous to mention								
AH47	3.2	4800	2	1 1/2	4.7		2	13 3/4
AH81	5.5	5000	2 1/2	1 5/8	7.98		2	13 3/4
AV47	2.2	3800	2	1 1/2	4.7		2	14 1/2
V55	5.5	3600	2 5/8	2 1/4	13.53		4	36 1/2
HR30	3.0	3600	2 5/16	1 13/16	7.61		4	28 1/4
HB30	3.0	3600	2 5/16	1 13/16	7.61		4	24

MCCULLOCH CORPORATION, Los Angeles, California
Principal Usage: Chain Saws

Model*	Rated hp.	R.p.m.	Bore	Stroke	Disp. Cu. In.	Comp. Ratio	2 or 4 Cycle	Weight
* Other models too numerous to mention								
1-40		6000	2 1/8	1 3/8	4.9	5.5/1	2	18
1-50		6000	2 1/8	1 3/8	4.9	5.5/1	2	18
1-70		7000	2 1/4	1 1/2	5.3	7.0/1	2	21
1-80		7000	2 1/8	1 1/2	5.3	7.5/1	2	25

BUYING CONSIDERATIONS

The person who is shopping for a gasoline engine will have no trouble in finding an engine for the job. There are many manufacturers of small gasoline engines and they produce engines in a variety of horsepower that are designed and engineered to satisfy every need. It is not uncommon to find a manufacturer's basic model with many variations to adapt the engine for a multitude of different jobs. Let us consider several points in buying an engine.

Cost: Don't be lead astray by bargain prices; true, a good bargain can be found now and then but the general rule, "You get what you pay for", is a good one. This is not to say that the most expensive engine is the best buy for you. Possibly the use the engine is to be put to does not require the added expense of special features, heavy-duty parts or more refined and embellished engine components. Your requirements might well be just an economical, dependable "workhorse" for cutting the grass once a week. However, if your engine will be exposed to continuous heavy usage it would be wise to buy a "heavy-duty" engine, one manufactured to take punishment for a long period of time without failure. The additional cost would be worthwhile in this case.

Reputation of the Manufacturer: Be certain the manufacturer's past record in the field justifies a faith in his new engines. Talk with engine dealers and individual owners of engines. Their opinions and experience may aid you with your selection.

Repair Parts: If you have an engine breakdown that requires a new part, will it be readily available? Can you "run down" to the local dealer and get the part or will you have to send to a factory that may be a thousand miles away? Delays caused by breakdowns can be costly as well as annoying. The availability of service and repair parts is an important consideration.

Power Requirements: Be certain the engine is big enough for the job. Constant overloading or constant operating at full throttle will shorten engine life. Talk with your dealer. He will help you select the correct power plant for the job.

Manufacturers today produce engines that operate on the four-cycle principle and engines that operate on the two-cycle principle; both have their place and both are entirely successful. Some engines are water cooled, some are air cooled, some have simple splash lubrication, some have pump lubrication, some have fuel pumps, some do not. In fact, no two engines look exactly alike or operate exactly alike.

Engine Warranties: Most engine manufacturers will give the purchaser of a new engine a warranty. A common type is good for ninety days. If, during the ninety days after purchase, any parts fail due to defective material or workmanship, the manufacturer will repair or replace the defective part.

In most cases the part or engine must be returned to the factory or to an authorized distributor. Most manufacturers require the owner to pay shipping charges to the factory. And, in some cases, the owner must also pay any labor costs involved in the repair.

If the engine has been damaged through misuse, negligence, or accident, most warranties are void. A few manufacturers require that you register your engine shortly after purchase; if you do not, the warranty is not valid. Generally, engine components such as magnetos, carburetors, starters, etc. are only covered by the terms of their individual manufacturer.

GENERAL STUDY QUESTIONS

1. If an engine can lift 150,000 pounds in three minutes, what is its horsepower?

2. Generally speaking, can larger horsepower engines do work faster than smaller horsepower engines?

3. What type of horsepower rating do most manufacturers use in advertising their engines?

4. What is a dynamometer?

5. In what way is SAE horsepower used?

6. Estimate the horsepower of a two-stroke cycle having a bore of 2″, one cylinder, a stroke of 1 1/2″, and operating at 3600 revolutions per minute.

7. What is engine torque?

8. What is volumetric efficiency?

9. What is the relationship between compression ratio and horsepower?

10. What is the relationship between piston displacement and horsepower?

11. List four important considerations for a person contemplating the purchase of an engine.

CLASS DISCUSSION TOPICS

- Discuss how horsepower increases as r.p.m. increases, remembering that engines deliver much less than their rated horsepower at slow speeds.

- Discuss and estimate the horsepower of a "live" engine in class. How does estimated horsepower compare with rated horsepower?

- Discuss how engine torque, volumetric efficiency, compression ratio, and piston displacement are functions of engine design.

- Discuss why there is a limit to the compression ratio that can practically be built into engines.

- Discuss the buying considerations for engines.

Unit 11

SAFETY AND POWER TECHNOLOGY

The maintenance, operation, repair and testing of the machines which are involved in a study of power technology requires that careful and constant emphasis be placed on safety. It is beyond the scope of this unit to deal with the particular safety considerations of each different prime mover and its vast number of applications. Instead, we shall discuss safety considerations in several common areas. In the units which deal with the various prime movers, we shall introduce specific safety information as it then applies. Fundamental areas of safety common to power technology applications include:

1. Safety Attitudes
2. Safety with Gasoline
3. Carbon Monoxide Safety
4. Safety with Basic Hand Tools
5. Safety with Basic Machines and Appliances
6. Safety with Small Gasoline Engines

SAFETY ATTITUDES

Safety is important to every student of power technology, both as an individual and as a member of a community. Your individual and personal safety is of prime importance to you — you want to stay alive and healthy as long as you can. The very fact that you are here today is testimony to the fact that you and others have cared about your safety. It is sometimes a bit frightening when you think of all the real dangers that a person must deal with throughout any ordinary day. Our machine age has surrounded us with a multitude of safety hazards. In transportation alone millions of power-driven vehicles spread out as a vast sea of hazards. Millions of industrial, construction, and scientific employees work with the tools of power technology every day. And this is not to mention the home itself which is also full of hazards — power appliances, household machines, fire, boiling liquids, electricity, and so forth. With these formidable hazards confronting us, it is necessary that emphasis on safety begin at the tender ages of early childhood and continue throughout a lifetime.

Accidents can be compiled and shown in statistical form and everyone is familiar with this practice. Accidents can be measured in terms of frequency, probability, medical costs, time lost from work, or in many other ways that serve a useful purpose in identifying hazards and in developing safety programs. But to the individual — his mind's eye often refuses to let him believe that he will ever be an "accident statistic."

On an individual basis, the statistical approach loses all its relevance — how can you put a number or a dollar sign on the personal cost and deep significance of human suffering when accidents occur! If you lose your hand, become blinded, maimed, or crippled you become an accident statistic but at that point you couldn't care less about statistics.

By nature we do not like to think of the unpleasantness of an accident happening to us even though we know that we are vulnerable as well as the next fellow. Fortunately, by our basic instincts, we seek to protect ourselves and those close to us from hazardous situations and accidents.

Power technology safety, like the highway, is a two-way street. We are concerned about ourselves, true — but our concern must go beyond ourselves to include the safety of all others. The person who says, "What I do is my own concern and doesn't affect others" is shortsighted and, in fact, he couldn't be more wrong.

Every person's actions and safety attitudes do have an effect on other individuals. Sometimes the effect is a slow environmental thing such as a parent's carelessness becoming a part of his children's attitudes and safety standards. Sometimes the effect on others is as direct as a headon collision on the highway caused by an irresponsible driver. The point is, in power safety you are responsible for yourself to be sure, but also you cannot escape your responsibility for the safety of others — you have these twin responsibilities.

A power technology student does not need a list of rules in his back pocket; what he really needs is an attitude in his mind that will prompt him to consider the safety aspect of every activity in which he engages. Most basic safety approaches to prime movers and machines are similar and he can transfer knowledge from one situation to·another if he has a responsible attitude. In power technology, safety is never sacrificed for the sake of speed or expediency. Rightfully, safety has priority over all else.

Your exposure to danger in power technology depends on your age and upon your activities. The safety requirements vary for all of us but generally we develop through stages.

Safety to the young child is largely a list of essential "do's" and "don'ts" — necessary rules that he follows in order to live safely in a complex environment that is full of hazards. His main job is to "stay out of the way." Neither his strength nor his judgment has matured to the point where he can operate machines or prime movers with any degree of safety to himself or others.

Teenagers are very susceptible to accidents. They have the physical size and strength to perform many tasks but, obviously, due to their age, they lack experience. How about judgment? Does a teenager have mature judgment that generally promotes safety? For some teenagers the answer is "yes" but on the other hand, we all know that for many teenagers the answer is "no." Some teenagers (like some adults) feel compelled to show off for others, to prove themselves, or use a machine to work off their frustrations at the expense of safety and good judgment. Psychologists list a wide variety of reasons why teenagers act as they do and most such reasons relate to the process of growing up and breaking the ties with home. But being young is no excuse for acting in a reckless manner that is dangerous for oneself or others. You can tell a great deal about a person's emotional maturity by watching him use prime movers and machines. Machines like the automobile can reveal an individual's entire safety code; how he regards himself and others.

The big danger for the adult is that he may become overconfident in his ability. He is strong, he has some experience behind him, and his judgment and emotional stability may be very good: in short he is an able, confident young man. Being in his prime he feels himself equal to and on top of every situation; his confidence may lead to carelessness and he, too, may become an accident statistic.

The adult on the job is the best accident risk. He has gained experience and his judgment has matured. Accidents to him take on a more serious meaning since he may have a family that depends on his ability to earn a living. Even though he is exposed to danger, he knows how to handle it and he wants to protect himself.

However, learning safety simply by maturing through age or by trial and error is not enough. The safety considerations of every power technology activity whether at work or at play must be studied. The defensive attitude of "If this happens, what might the result be." must be woven into the fabric that makes up a person's overall judgment. To be sure, some "safety" is plain commonsense but many safety requirements can and should be learned from books, magazine articles, machinery instruction booklets, and repair manuals. In a technical age you can't go the entire distance on just commonsense; you need specific information that pertains to the particular prime mover and its application.

SAFETY WITH GASOLINE

Understanding the nature of gasoline and its safe use is much more important today than it was 15 to 20 years ago. In days past, the average person's use of gasoline was restricted to the family automobile. Gasoline was pumped into the tank by the service station operator and the purchaser never really got close to it.

Today, however, in millions of modern households this fluid has become a common everyday item. The presence of this liquid, which packs more power than TNT, is readily accepted and really alarms no one. In fact, many persons have become frighteningly casual with this extremely volatile and explosive liquid.

Fig. 11-1 Store Gasoline and other Flammable
Liquids in their Proper Containers.
Courtesy: National Safety Council.

Gasoline is around the home today (in the garage, tool shed, or barn) largely because the use of small gasoline engines has become so widespread. Lawnmowers, garden tractors, snow throwers, go-carts, utility vehicles, motor bikes, outboard motors, and chain saws need gasoline for operation and, as a result, a gasoline can is now found in nearly all households. And accidents do happen; it is estimated that several hundred persons lose their lives every year due to accidents involving gasoline and other flammable liquids. In addition, several million dollars of property damage is caused by such accidents.

The volatility of gasoline is the characteristic that makes it such an ideal fuel but herein also lies the prime danger. Gasoline can vaporize and explode in the atmosphere just as it can in the engine. A concentration of 1 to 7 cubic feet of gasoline vapor per 100 cubic feet of air represents a flammable condition. Any spark or flame can touch off gasoline as a fire or explosion.

Gasoline should not be stored inside the home — it is just too dangerous a liquid. It should be kept in a garage, tool house, or some other outside building. Storage in the basement is bad; a gas can could be tipped over by a pet or small child and if the seal were imperfect, gasoline would trickle out and begin to vaporize. Given the right chain of circumstances, and there are many on record, the whole basement could blow up — triggered by the flame from a furnace or water heater or from the arc of an electric switch.

More and more states are recognizing the hazards of gasoline around the home and are developing safety regulations that will help protect the public. Storage containers is one important area. Ideal storage is in a red metal can clearly labeled GASOLINE. Red is a universal color indicating danger. A metal can will not break if it is accidentally dropped or struck a blow.

Follow these rules for the storage of gasoline:

1. Store gasoline in a metal container.

2. Store gasoline in an outbuilding — not inside the home.

3. Do not store gasoline during the "off season." It is too dangerous a liquid just to "have around."

4. Do not allow small children to have access to the gasoline supply.

5. Store gasoline away from flames, excessive heat, sparks caused by static electricity, sparks caused by electrical contacts, or sparks caused by mechanical contact.

6. Have the proper portable fire extinguisher readily available for use as needed.

Using gasoline safely involves decisions and judgments. It should be regarded as dangerous. Further, it should be regarded as fuel for gasoline engines and not as a handy all-purpose solvent or cleaner or fire starter. Again, the volatility and explosive nature of gasoline makes it unsuitable for any type of general household use. Gasoline used as a cleaning agent for paint brushes or to cut grease sets up real safety hazards. Accidental ignition can occur while the gasoline is being used, and after it has been dirtied, the problem of disposing of the liquid may be hazardous.

As a fire starter, it has been proved time and again that a blazing trail of gasoline can outrun any person. Never use gasoline as a fire starter. The result may be a fire or explosion that encompasses both the proposed fire and the person with the gasoline can.

Observe these precautions when using gasoline:

1. Use gasoline only as a fuel for engines or devices where its use is clearly intended.

2. Always regard gasoline as dangerous.

3. Never smoke cigarettes, pipes, or cigars around gasoline.

4. Never fill the tank of a hot engine or a running engine if the tank and engine are at all close to each other. Gasoline splashed on a hot engine may ignite or explode.

5. When pouring gasoline from container to tank, reduce the possibility of a static electricity spark by having metal-to-metal contact. A metal spout held against the tank opening is good or a funnel that contacts both container and tank is good. Pouring gasoline through a chamois may present a static electricity hazard.

Fires are an unpleasant thought but an understanding of fires and fire extinguishers may prove to be very useful information at some later date. Most fires fall into one of these categories:

Class A - Wood, cloth, paper, rubbish

Class B - Oil, gasoline, grease, paint

Class C - Electrical equipment

Class A fires naturally are the most common. Extinguishing these fires consists of quenching the burning material and reducing the temperature below that of combustion. Some extinguishing methods also have a smothering effect on the fire. Most persons are familiar with the soda acid extinguisher that is seen in many public buildings, industries, and places of business. To actuate this extinguisher it is simply up-ended, the water stream that is expelled should be directed at the base of the fire first, then back and forth following the flames upward. These extinguishers are not suitable for Class B fires such as gasoline fires. Foam fire extinguishers produce a foaming liquid when the extinguisher is inverted. For Class A fires, the foam should be directed at the base of the flames.

Fig. 11-2 Have Fire Extinguishers Readily Available and Know the Correct Use of the Various Types.
Courtesy: National Safety Council

The control and extinguishment of Class B fires, which includes gasoline, presents a far greater hazard than Class A fires do and careless or unknowing action against this fire can place the firefighter in extreme danger. Basically, the technique is to cut off the oxygen supply which feeds the burning liquid or to interrupt the flame.

Carbon Dioxide (CO_2) extinguishers contain this gas under high pressure. Carbon dioxide will not support combustion and therefore has the effect of cutting off the oxygen in the air and smothering the flame. The extinguisher should be used with a slow sweeping action that travels from side to side, working to the back of the flame area. The discharge horn of the extinguisher becomes extremely cold during discharge and it should not be touched. Another hazard is that in a small room the extinguisher may produce an oxygen-short atmosphere that is dangerous to the firefighter himself.

Foam fire extinguishers produce a water-base foam that can smother the fire. In using an extinguisher of this type on Class B fires, direct the foam at the back of the fire allowing the foam to spread onto the flame area. This minimizes the possibility of splashing the flaming liquid of an open container fire.

Dry chemical extinguishers are excellent for Class B fires and for Class C fires. When the dry chemical is expelled under pressure, it interrupts the chemical flame chain reaction and thereby extinguishes the fire.

Class C fires involve electrical equipment and therefore water-base extinguishing materials are not suitable. Water on or around an electrical fire creates a shock hazard. If the electrical energy can be completely and positively shut off, the burning material can be handled according to its nature. If the equipment is energized, the fire must be fought with CO_2 extinguishers, dry chemical extinguishers, or vaporizing liquid extinguishers.

CARBON MONOXIDE SAFETY

The thought of a poison that is colorless, odorless, and tasteless is a bit sobering; it sounds evil and diabolical. A poison with dangerous characteristics like these can conjure up morbid thoughts suitable for chemical warfare or horror movies; yet we live with a poison just like this every day — its name: carbon monoxide.

The hazard of carbon monoxide is real. Carbon monoxide is a killer and, in fact, it is responsible for more poisoning deaths than any other deadly poison.

This gas is the result of the incomplete combustion of solid, liquid, or gaseous fuels of a carbonaceous nature. At home the gas is found along with improperly adjusted hot water heaters. In industry carbon monoxide can be found with kilns, oven stoves, foundries, smelters, mines, forges, and in the distillation of coal and wood, to list a few. However, it also exists in a much more common circumstance — the exhaust gases of internal combustion engines such as those used on the automobile. When you operate an automobile you are producing the deadly poison, carbon monoxide. Knowledge about this poison and how to eliminate or minimize the danger is an important topic for everyone.

The action of the poison on the human body can be quite rapid. The hemoglobin of the blood has a great affinity for CO, 300 times greater than that for oxygen. When the CO is combined with the hemoglobin it has the effect of reducing the amount of hemoglobin available to carry oxygen to the body tissues. If large quantities of CO combine with the hemoglobin, the body becomes starved for oxygen and literally suffocates.

Ventilation is of prime importance in preventing carbon monoxide poisoning. Unburned gases and exhausts must be carried away as effectively as possible by using chimneys, ventilation systems, and exhaust systems. These systems must be efficient because even small amounts of the gas can cause a dulling of the senses that spells danger.

Exhaust gases and carbon monoxide can get inside an automobile in several ways, through a defective exhaust system — tail pipe, muffler, or manifold, or through rusted-out or defective floor panels. And, if there is excessive CO around the automobile due to faulty exhausting or a poorly tuned engine, it

CREEPING KILLER - CO

Fig. 11-3 Do not Underestimate the Danger of Carbon Monoxide.
Courtesy: National Safety Council.

can come in right through an open window. For example, an open tailgate window on a station wagon can produce a circulation pattern that brings exhaust into the auto through the tailgate window. In the passenger section the carbon monoxide can build up to dangerous proportions. Station wagon rear windows should be closed, especially if other windows are open.

Consider the front window vents of the auto. When these are opened only slightly they draw air from the passenger compartment and if there is a hole at some exhaust concentration outside the compartment, exhaust can be drawn in. Generally, it is best to have the front vent windows open plus the side windows rolled down a little bit in order to provide an adequate flow of fresh air. The air ducts that bring air in through the heater system are also a good ventilation source.

Knowing the symptoms of carbon monoxide poisoning and heeding their warnings could someday be a life-and-death matter. These symptoms include a tightness across the forehead, followed by throbbing temples, weariness, weakness, headache, dizziness, nausea, decrease in muscle control and an increased pulse rate and increased rate of respiration. If you feel any of these symptoms, stop the car or, if at home or at work, stop the activity and get some fresh air — you may be being poisoned by carbon monoxide.

If you discover a person who has passed out from carbon monoxide poisoning, remove him to the fresh air immediately and if breathing has stopped or if the victim is only gasping occasionally, begin artificial respiration at once. Also, have someone call a doctor and/or a fire or emergency squad.

To prevent carbon monoxide poisoning, follow these rules:

1. Do not drive an auto with all the windows closed.

2. Do not operate internal combustion engines in closed spaces such as garages or small rooms, unless the room is equipped with an exhaust system.

3. Keep carbon monoxide producing engines and devices tuned or adjusted properly in order to reduce the output CO gas.

4. Do not sit in parked cars with the engine running.

5. Be mindful of the importance of properly working ventilation and exhaust systems.

SAFETY WITH BASIC HAND TOOLS

It may be the space age with many sophisticated prime movers and it may also be the age of automation for business and industry, but for all the advances in technology the common hand tool is around as strong as ever. Long ago, common hand tools forged out the machine age; the machine age fashioned space age hardware of today. But still, hammers, pliers, wrenches, screwdrivers, saws, chisels and their humble associates, serve the home mechanic, the professional mechanic, and the research technician with equal effectiveness.

The vision of the old craftsman cherishing his tools is a good one to keep and to emulate. Tools are our friends; they extend our ability to work with our hands and amplify our strength. Tools should be treated with respect and well cared for. Tools should be stored in a fashion that will protect their vital surfaces, cutting edges, true surfaces, etc. A wall tool panel with individual hangers or holding devices for each tool is ideal since the tools are easily visible, quickly accessible and nicely protected. Tools thrown in a drawer in a haphazard manner will scratch, dull, or otherwise damage each other and are a hazard to the person who is rummaging around for the right tool.

When tools must be carried in a toolbox they should have individual compartments if possible. Lacking this, the more sensitive tools and tools with cutting edges should be wrapped in cloth.

Moisture and its accompanying rusting is a problem; so, if tools get wet, wipe them dry before storage. Also, the rusting problem can be combatted by wiping the tools with a slightly oily rag. Should a tool pick up dirt during use, wipe it clean before putting it away.

Fig. 11-4   All "Struck" Tools Should be Properly
Dressed on the End.
Courtesy: General Motors Corp.

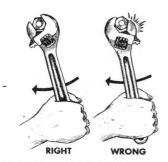

Fig. 11-5   The Hammer Face Should Strike the
Work with its Full, Flat Face.
Courtesy: General Motors Corp.

Fig. 11-6   Use an Adjustable Wrench so the
Strain is on the Fixed Jaw.
Courtesy: General Motors Corp.

Fig. 11-7   Pliers Inevitably Scar the Flats of a Nut.
Courtesy: General Motors Corp.

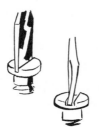

Fig. 11-8   Use the Correct Size Screwdriver for the
Job.   Courtesy: National Safety Council.

The cardinal rules for the safe use of hand tools are: (1) Use tools that are in good condition and well sharpened. The dull tool is not only inefficient but it is more likely to slip during use. (2) Use the correct tool for the job and use it as it is intended to be used. The incorrect use of tools has been the cause of many accidents.

A metal chip in the eye is not a pleasant thought and, in some operations, hand tools can present just this hazard. It is not difficult to convince power machine operators of flying chip hazards but many persons are hard to convince that dangerous little projectiles can come from the ordinary use of simple hand tools. Just think of a mis-struck nail ricocheting across the room and you will agree that even a hammer can be dangerous. Especially dangerous are chipping operations with a cold chisel. All struck tools such as chisels and punches can mushroom out at the end over a period of use. These ends should be dressed on a grinder to eliminate the possibility of tool chips breaking and flying from the tool when it is hit. Hammers themselves may chip if they are struck side or glancing blows — hammers should strike the work with their full face.

At first thought, a wrench seems to be just about as safe a tool as you can find and, if it is used properly, it is. Used incorrectly, the wrench can be dangerous. Be certain to use the exact size wrench for the nut or bolt to prevent slipping or damage to the flats of the nut or bolt. If an adjustable wrench is used, snug it up tight against the flats of the nut or bolt and be sure that your direction of force will place the major strain on the fixed jaw and not on the movable jaw. Most of all, pull a wrench toward you, do not push it. Pushing can be dangerous; the nut or bolt may suddenly loosen or the tool may slip and your hand may go crashing into a steel part.

Screwdrivers are probably the most misused of all common tools; the list of abuses ranges from using it to open paint cans to using it as a wrenching bar. The tool is simply intended to be used in tightening or loosening various types of screws and therefore its tip must be preserved for this purpose. Use a screwdriver that is the correct size. It should fit the screw slot snugly in width and the blade should be as wide as the slot is long. The novice will often try to use one medium size screwdriver for all jobs. Have several sizes of screwdrivers in your tool assortment. It is a poor practice to hold the work in one hand in such a manner that a slip would send the screwdriver into the palm of the hand. Screwdrivers, pliers, or other tools that are used for electrical or electronic work should have insulated handles.

Files are rarely sold with handles attached but they should be so equipped. The tang is fairly sharp and would cut or puncture the hand if the file suddenly hit an obstruction. The brittle nature of files makes them a poor risk and a dangerous risk as a pry bar; do not use them for this purpose.

SAFETY WITH BASIC MACHINES AND APPLIANCES

Safe use of basic household machines, mechanisms, and appliances during normal use, adjustment, and repair is important to all. Reference is made to power hand tools; power tools; appliances such as stoves, refrigerators, dishwashers, clothes driers, washing machines, air conditioners, electronic equipment such as television sets, record players, and radios; powered hobby or sports equipment; small home appliances for kitchen or personal use; and so forth. Several common rules can be developed for these many machines.

1. Before attempting the adjustment or repair of a machine, always unplug or disconnect the machine or device from the electrical power source. Be certain that someone else will not accidentally reconnect or plug in the machine while you are working on it. If the power source is in a remote location, tag the source "Do not connect. Equipment under repair."

2. Before attempting the adjustment or repair of machines driven by engines, stop the engine, disengage the clutch, and remove the spark plug lead to eliminate the possibility of accidentally restarting the engine.

Fig. 11-9 Tools Used in Electrical and Electronic Work Should Have Insulated Handles. Courtesy: National Safety Council.

3. Do not neglect to ground electrical equipment that requires grounding. Use the three-prong plug on portable tools or devices and properly ground the convenience outlet.

4. Keep loose clothing away from rotating parts that could "grab" the cloth and thereby pull you into the machine.

5. Protect your eyes with safety glasses if there is any possible chance of flying chips or breakage.

6. Thoroughly guard all belts, pulleys, chains, gears, etc. Do not remove these guards or allow them to be removed by others.

7. Study the instructions that come with the machine and be certain that you understand the operating principles and the safety precautions before operating or repairing the machine.

8. Use the machine only in the manner and for the purpose for which it was designed.

9. During adjustment and repair, retighten all nuts, bolts, and screws securely.

10. Unless you are qualified and understand the danger, keep away from high voltage areas in electronics equipment.

SAFETY WITH SMALL GAS ENGINES

The application of small gasoline engines to a multitude of common home chores has produced a safety problem that is significant. Safety education about lawnmowers, chain saws, sports vehicles, outboard motors, and the like, should begin at an early age, even before a person has reached the age and emotional maturity that qualifies him as an operator. These machines are not as formidable in appearance as, for instance, an automobile, yet they pack tremendous power that can easily cut or maim the human body.

Small engine applications are often a person's initial exposure to gasoline engines and powered machinery where he is the actual operator. There is danger in the fuel and its use around the machine, and there is danger in the use of the machine itself. Unfortunately, some adults assign small engine tasks or permit the use of small engines in recreational activities when the young operator possesses neither the physical readiness nor the maturity of judgment that is necessary for safe operation. And sometimes the preparation or instruction given the operator includes only starting the engine and stopping the engine with no attention to the safety considerations that are necessary. Trial and error is a very poor way to learn the use of any prime mover or machine.

General safety rules for all small engine applications:

1. Do not make any adjustments or repairs to machinery being driven by an engine without first stopping the engine and removing the high tension lead from the spark plug. It is possible to accidentally turn the engine over and restart it especially if the engine is warm. With the spark plug cable removed from the spark plug, the engine cannot start.

2. Do not fill the gas tank when the engine is running or hot. If gasoline is spilled on a hot engine, a fire or explosion can result. If you have not accomplished your job and the engine needs more gasoline, stop the engine and let it and yourself cool off before refilling the tank. It is a good idea to start each job with a full gas tank.

3. Do not operate a gasoline engine in a closed building due to the carbon monoxide hazard.

4. Read the equipment instruction book. Know and understand the equipment before operating it.

5. Keep the equipment in perfect operating condition with all guards in place.

The power lawnmower is a typical example of a small engine application and probably the most widespread in its use; millions are in use. There use is not restricted to one age group; people of all ages regard the power lawnmower as a necessary friend. However, it never pays to become too trusting, friendly or familiar with this or any other machine.

On lawnmowers, the cutting blade is a real menace. The whirling blade of a rotary lawnmower can cut through large sticks and pick up and hurl foreign objects like projectiles. Your body could be cut by the blade as easily as if it were a small twig. Studies show that about half of the accidents that occur with rotary power lawnmowers are caused by thrown objects, and most accidents that involve the operator himself result in injury to toes, fingers and legs.

The person who operates a power lawnmower accepts a safety responsibility for both himself and for others. These safety rules should be followed:

1. When starting a lawnmower, be certain that you are standing clear with feet and hands away from the blade.

2. Do not mow wet grass — it can be slippery. In the case of electric lawnmowers, wet grass can present an electric shock hazard.

3. Keep away from the grass discharge chute; sticks, stones, etc. will most likely be thrown out of this opening.

4. When mowing hills or inclines, do not mow up and down; mow across the face of the slope.

5. Do not pull a lawnmower — push it. If you fall while pulling you may pull the mower right onto yourself.

6. Inspect the lawn for rocks, sticks, wire, and other foreign objects that can be converted into lethal projectiles by the mower blade.

7. Allow only a responsible person to operate a power mower. Small children are naturally attracted to this type of activity. It is your responsibility to see to it that they are completely out of the mowing area — preferably in the house.

8. Do not use a riding lawnmower as a play vehicle.

9. Keep self-propelled lawnmowers under full control. Don't let a lawnmower control you or pull you around the yard; you must have full command of the situation at all times.

10. Never leave a mower running unattended.

GENERAL STUDY QUESTIONS

1. Explain why only practicing "commonsense" precautions is not a completely adequate approach for a power technology student.

2. Prepare a list of prime movers that a typical teenager might be expected to operate safely.

3. Why is the young adult often susceptible to accidents?

4. What characteristic of gasoline makes it such a dangerous liquid?

5. List several requirements for gasoline storage and precautions for gasoline use.

6. What type of extinguishers are best for putting out gasoline fires?

7. How is carbon monoxide produced? What is its effect on the body?

8. Explain why it is important to disconnect electrical devices from their electrical source before repairs or adjustments are made.

9. Summarize the hazards in using hand tools.

10. List several safety rules for operating power lawnmowers.

CLASS DISCUSSION TOPICS

● Discuss how safety becomes a part of a person's personality.

● Discuss how every individual's safety attitudes can affect other persons.

● Discuss why learning safety through "trial and error" is a poor practice.

● Discuss the proper use and storage of gasoline.

● Discuss the various ways to prevent carbon monoxide poisoning.

● Discuss equipment instruction books and their importance to safe machine operation.

● Discuss the dangers of loose clothing to mechanics, technicians, or machine operators.

● Discuss the proper care and storage of hand tools.

Part II Other Internal Combustion Engines

Unit 12

THE AUTOMOBILE ENGINE

The automobile engine is a big brother to the small gasoline engine. Generally, the engines have the same basic parts, the same basic systems, and often operate on the same four-stroke cycle principle. Of course, every engine is designed and engineered for a specific purpose but the basic parts and operating principles remain virtually unchanged. A thorough-going knowledge of any one engine puts a person just a few steps from understanding other engines.

The backbone of the engine is the block. The cylinders are cast together: one casting contains all the cylinders, usually six or eight. Six-cylinder engines usually have their cylinders in a straight line (in-line engine) while eight-cylinder engines have a "V" block. The V-8 has four cylinders on each side and the block takes the "Y" shape. The lower part of the block is called the crankcase and provides mounting for the crankshaft and space for the crankshaft to revolve.

Below and bolted to the block is the oil pan. This part is a stamped-out pan that acts

as a reservoir for the engine's oil supply. On small engines this part might be called the sump, base, or crankcase.

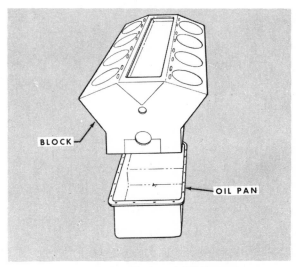

Fig. 12-2   Block and Oil Pan.

The cylinder head is bolted to the top of the block, one head for six-cylinder engines, two heads for V-8 engines. Most modern engines have overhead valves or valves mounted in the cylinder head (valve-in-head). On most small engines the valves will be found in the engine block. The valve system and cylinder head are necessarily more complicated on overhead valve engines. Instead of simply pushing the valve up for opening, the valves must be pushed down for opening; more parts are needed. Rocker arms mounted on a shaft make this change in direction. The head, therefore, contains the intake and exhaust valves, rocker arm shaft, rocker arms, valve springs, and other valve parts. The camshaft is in the engine block. The cam movement is transferred to the rocker arm through a push rod. The train is: camshaft, tappet, push rod, rocker arm, valve. When the nose of the cam pushes up on the tappet, push rod, and rocker arm, the rocker arm also pushes the valve down (open).

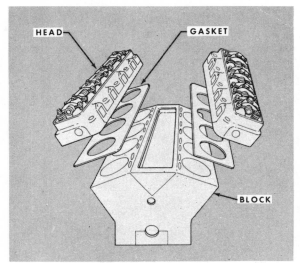

Fig. 12-1   Block, Head, and Gasket of a V-8 Engine.

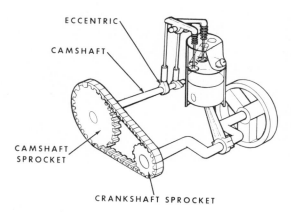

Fig. 12-3   Operation of the Rocker Arm.

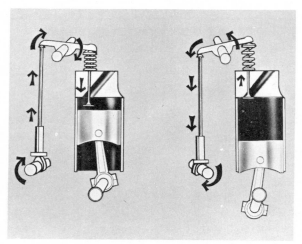

Fig. 12-4   Valve-in-head Arrangement.

Fig. 12-5   Valve-in-head Arrangement of a Six-cylinder Engine.

Correct valve timing is vital to engine performance. The cams must open and close the valves at the correct instant. The crankshaft and camshaft must, therefore, be perfectly coordinated. Timing marks on the gears indicate the proper alignment. Incorrect assembly by an inexperienced mechanic would be the most probable source of trouble in valve timing. Many engines use a timing chain between the crankshaft and camshaft gears. In this case, if the timing chain becomes worn, improper timing and sluggish engine operation can result.

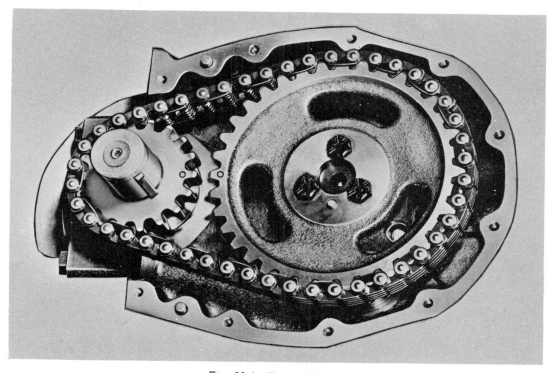

Fig. 12-6   Timing Chain.

Pistons are usually made of an aluminum alloy to reduce weight. The piston is equipped with cast iron piston rings, compression rings to reduce power loss through blow-by and oil rings to scrape away excess oil and spread out an even oil film on the cylinder walls. Of course, the piston is linked to the crankshaft by the wrist pin, connecting rod, and connecting rod cap.

The crankshaft converts the reciprocating motion of the piston to the necessary rotary motion. The six-cylinder crankshaft has six crank throws, the offset sections, evenly spaced around the axis of rotation to deliver smooth power. The V-8 engine has four crank throws spaced around its axis, one throw for each two opposite cylinders. When one piston is down, the opposite piston is up. Two connecting rods, therefore, are fastened to each throw. The crankshaft has journals which rest in the main bearings of the crankcase.

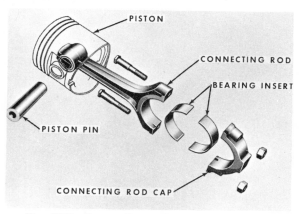

Fig. 12-7   Piston, Piston Pin, Connecting Rod, Bearing Insert, and Connecting Rod Cap.

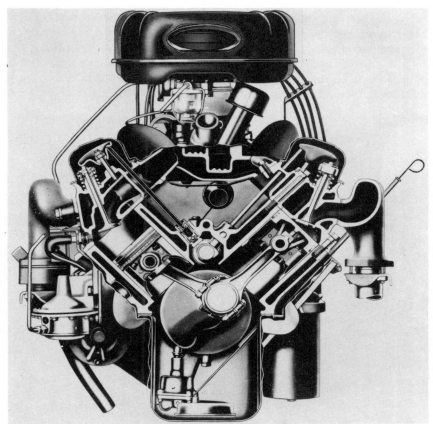

Fig. 12-8   Engine Cross-section Across Crankshaft.

Fig: 12-9   Engine Cross-section Along Crankshaft.

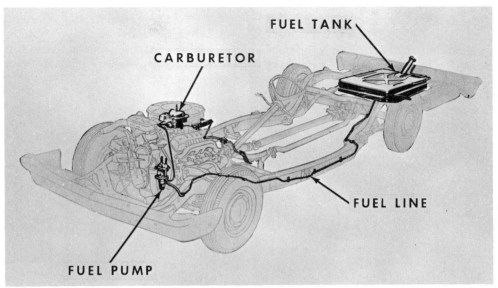

Fig. 12-10 Fuel System.

The fuel system is made up of the fuel tank, fuel line, fuel pump, and carburetor. These same essentials are also found on most small engines. The tank is vented to the atmosphere to prevent a vacuum from forming inside the tank and the tank is equipped with a filter to strain out foreign material.

The fuel pump is operated by a cam located on the camshaft. The pump brings gasoline from the tank, delivering it to the carburetor. The main parts of the pump are: body, sediment bowl, diaphragm, pump rocker arm, inlet valve, and outlet valve. Cam action moves the diaphragm in and out, at one time forcing fuel out through the outlet valve and on to the carburetor, then drawing fuel into the pump from the gas tank. In actual operation the pump delivers fuel only when the carburetor calls for it; that is, when the fuel inlet valve in the carburetor is open. If the float chamber is full and the fuel inlet valve on the carburetor is closed, the pump will not deliver fuel. Even though

the pump rocker arm continues to move, the diaphragm will remain steady until more fuel is demanded.

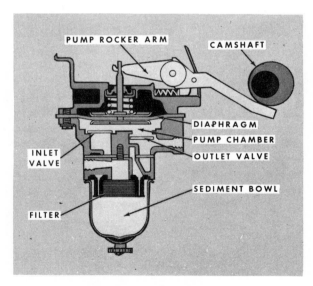

Fig. 12-11 Fuel Pump.

Automobile carburetors are generally more complex than small engine carburetors. The operating principles are the same for each but the larger carburetors have more refined systems. In either case, the carburetor's job remains the same: to mix air and gasoline in the proper proportions and to deliver the

mixture to the combustion chambers. A typical carburetor would consist of (1) idle system, (2) main system, (3) power system, and (4) accelerating system. In discussing the systems, it is assumed that the reader already has gained an understanding of carburetor fundamentals.

The idle system supplies additional fuel for idle or slow speeds. When a relatively small amount of air is drawn through the venturi section, the venturi pressures are not very low. Therefore, little fuel from the main discharge system enters the venturi section to mix with the air. The mixture would be too lean without the addition of the idle system.

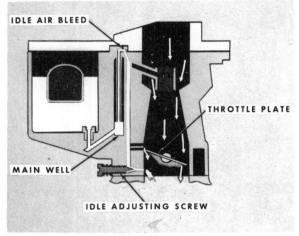

Fig. 12-12 Idle System.

Fuel from the main well travels up the idle tube and down to the idle jet during idle speeds. Idle discharge jets are located just behind the throttle plate and with the throttle closed down for idle speed, the pressure behind the throttle is very low. Atmospheric pressure forces the gasoline to this low pressure area. As the throttle is opened and the engine speeds up, the fuel supply shifts from the idle system to the main system.

During normal driving speeds the gasoline is delivered by the main fuel system. Low pressures are attained in the venturi section and atmospheric pressure pushes the fuel into the airstream.

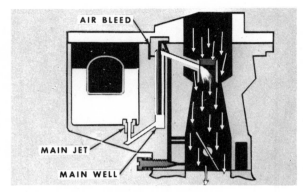

Fig. 12-13 Main System.

The power system supplies extra fuel for the richer mixture necessary for heavy loads or high speeds. A valve in the bottom of the float bowl opens to allow additional fuel to enter the main well. The valve is operated by intake manifold pressure. At low speeds (throttle closed) intake manifold pressure is very low and the spring-loaded power valve is held closed. However, at high speeds (throttle opened wide), the intake manifold pressure is relatively high, enabling the spring tension to open the power valve, admitting extra gasoline. The power system will stay in operation until the throttle is closed down some or the load is lightened.

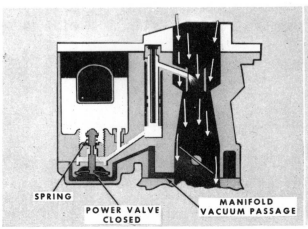

Fig. 12-14 Power System.

The <u>acceleration system</u> operates when the throttle is opened. Because good acceleration requires a richer fuel mixture, the system sprays additional gasoline into the carburetor. The acceleration system shown in Fig. 12-15 has a diaphragm rod between the throttle plate and the diaphragm. When the throttle is opened, its motion is transferred to the diaphragm, moving the diaphragm inward. The fuel in the diaphragm chamber is then forcefully sprayed into the venturi section of the carburetor.

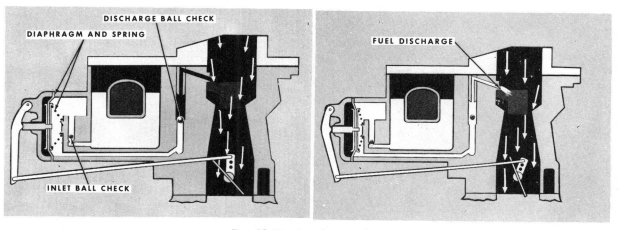

Fig. 12-15   Acceleration System.

Most carburetors used on automobiles are equipped with an automatic choke. Unlike the manual choke, the automatic choke cannot be adjusted by the driver. Bear in mind that the function of the choke (automatic or manual) is to provide a rich fuel mixture for starting the engine. When the engine is first started, the choke is in its closed position. As the engine warms up, the choke automatically moves into its open position. The movement of the choke from its closed position to its open position is caused by a thermostatic spring. When the spring is cold, it holds the choke in its closed position. As the spring begins to warm up, it appears to unwind a bit. This unwinding moves the choke from its closed position to its open position. Heat is supplied to the spring from the exhaust manifold.

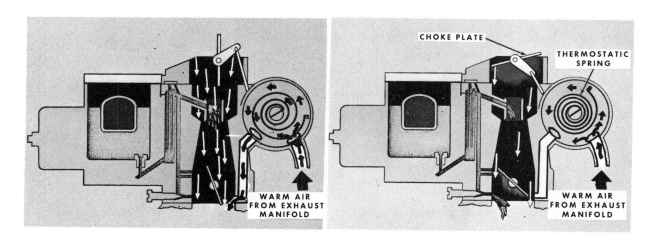

Fig. 12-16   Automatic Choke.

The exhaust system consists of the exhaust manifold, inlet pipes, muffler, and tailpipe. The muffler is designed to reduce noise without letting the exhaust pressure build up to a point that would be harmful to engine efficiency. The manifolds carry the exhaust gases from the engine to the inlet pipes. The gases pass through the inlet pipes to the muffler. After passing through the muffler, the gases are discharged through the tailpipe.

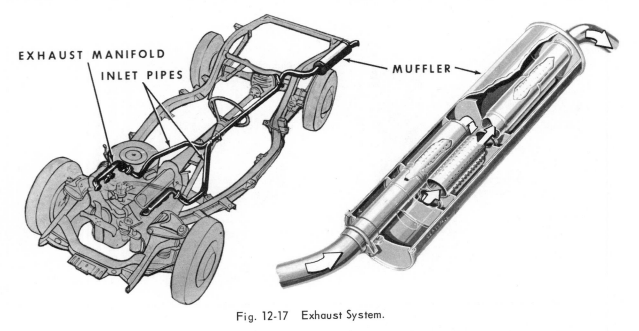

EXHAUST MANIFOLD
INLET PIPES
MUFFLER

Fig. 12-17   Exhaust System.

IGNITION SYSTEM

The automobile ignition system is very similar to the small engine magneto ignition system in many respects. Many of the parts used in the two systems are the same; both share a common electrical theory of operation.

Briefly, the parts of the system are:

1. Battery: Provides the energy source for the primary circuit.

2. Generator: Keeps the storage battery charged.

3. Switch: Opens and closes the electrical circuit to stop and start ignition.

4. Ignition Coil: Consists of a primary and a secondary coil. The primary coil contains relatively few turns of wire and produces relatively low voltages. The secondary coil contains thousands of turns of wire and produces very high voltages.

5. Distributor: Delivers high-tension voltage to the correct spark plug at the correct time.

    a. Distributor Rotor: Spins around, moving across the contact points for the spark plug leads.

    b. Breaker Points: Act as a switch in the primary circuit.

    c. Condenser: Protects the breaker points by acting as an electrical storage tank.

    d. Breaker Cam: Opens and closes the breaker points.

6. Spark Plug: Ignites the fuel mixture in the cylinder as high-tension voltage jumps the spark gap.

7. Spark Plug Leads: Carry high-tension voltage from the distributor to the spark plugs.

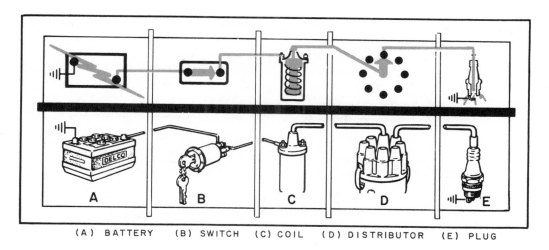

(A) BATTERY   (B) SWITCH   (C) COIL   (D) DISTRIBUTOR   (E) PLUG

Fig. 12-18   Ignition System.

A review of the electrical theory of the ignition system will be helpful in understanding its operation. When the distributor breaker points are closed and the ignition switch is on, six or twelve volts of current travels from the battery through the ignition coil low voltage circuit, then through the closed breaker points and back to the battery through ground or a return wire. This is known as the primary circuit.

With the flow of electrons in the primary circuit, energy is stored in the coil in the form of a magnetic field. As the circuit breaker cam rotates, it opens the breaker points just as a piston is reaching the top of its compression stroke and as the distributor rotor is passing the contact point for that cylinder. When the breaker points open, current immediately stops and the magnetic field that was built up col-

lapses. This collapse of the magnetic field transforms the energy in the coil into a high voltage surge sufficient to create a spark at the spark plug. This voltage is usually between 15,000 and 20,000 volts. The high voltage is created in the secondary circuit and follows this path: high-tension wiring, distributor rotor and contact points, spark plug lead, and spark plug.

The breaker cam continues to rotate in time with the rotor, delivering high-tension voltages to the spark plugs at the correct time. Each time the breaker points open, a spark occurs.

The spark must be timed to occur as the piston reaches the top of its compression stroke. Engine speed and load determine the exact time that the spark should occur. A centrifugal spark advance mechanism and a vacuum spark advance mechanism accomplish this task.

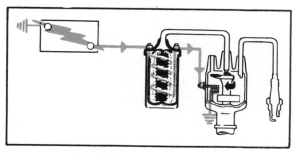

(A) ENERGY STORED AS
MAGNETIC FIELD

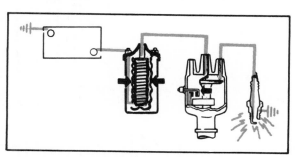

(B) WHEN CONTACTS OPEN, COIL
PRODUCES HIGH VOLTAGES

Fig. 12-19   Primary Circuit.

In automobiles where a 12-volt electrical system is used, a modification of the simplified system is sometimes made by using a resistor which has a special contact. This special contact allows a parallel circuit to be set up which bypasses the resistor during the cranking interval. This contact may be installed in the solenoid switch or in the ignition switch. By using this bypass circuit, full battery voltage is supplied to the coil to insure quick starting.

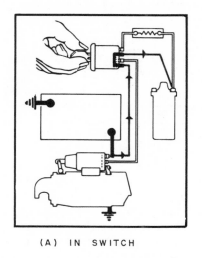
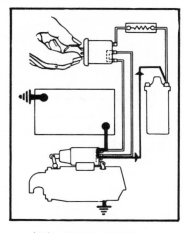

(A) IN SWITCH   (B) IN SOLENOID

Fig. 12-20 Resistor Bypass During Starting.

Some 12-volt systems use an ignition coil with an external series resistor assembly, the resistance of which is unaffected by change in temperature. This special resistor dissipates heat which tends to form in the coil and also reduces the effect of low temperatures on the primary winding. With the resistor connected in series, the primary winding of the ignition coil represents about half of the primary circuit resistance and is the only part strongly affected by temperature. Since only part of the total resistance is affected, there is less current increase in cold weather, and the system, therefore, is much less subjected to blued or oxidized distributor contact points than a system without a resistor.

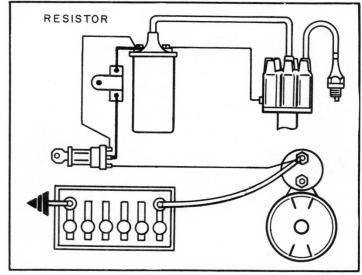

Fig. 12-21 External Series Resistor Circuit.

LUBRICATION

Just as in the small gasoline engine, the chief function of the lubrication system is to reduce friction. Friction causes damaging heat and excessive wear of engine parts. Lubricating oils also perform the auxiliary jobs of providing a better power seal, cleaning by washing away small bits of metal and dirt, and cooling by improving heat transfer between parts.

An oil pump moves the oil throughout the engine. Generally, the oil from the pump goes to the oil filter; then through passages drilled in the engine block to the crankshaft and camshaft. Also, oil is sent to the rocker arm shaft where it lubricates the rocker arms, valve stems, pushrods, and, as it drains back to the sump, the timing sprockets and chain and distributor drive gear.

Oil to the crankshaft, camshaft, rocker arm shaft, connecting rod bearings, main bearings and other necessary parts is delivered through small drilled passages within the parts themselves; this is usually referred to as full pressure lubrication.

In addition, cylinder walls and the parts inside the crankcase are splashed or sprayed with oil. Whether the part requires a continuous oil bath or a moderate one, the system is designed to meet the specific needs of all parts.

A gear type oil pump is capable of delivering more oil than the system can distribute. To prevent excessive pressures from building up in the lubrication system, an oil pressure relief valve is installed. When the maximum desired pressure is passed, the spring loaded relief valve will open allowing the excess oil to drain back to the crankcase and sump, thereby preventing pressure buildup.

The oil filter continually cleans the lubricating oil; dirt, soot, minute bits of metal, and any foreign matter is trapped in the filter. The filter may be made of cotton waste, cellulose, inert earth, paper, or any material that will effectively filter the oil.

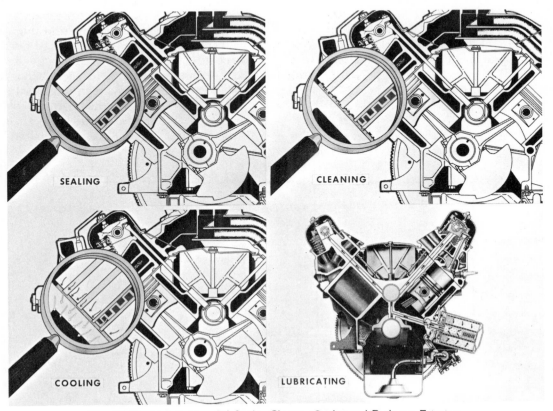

Fig. 12-22 Lubricating Oil Seals, Cleans, Cools, and Reduces Friction.

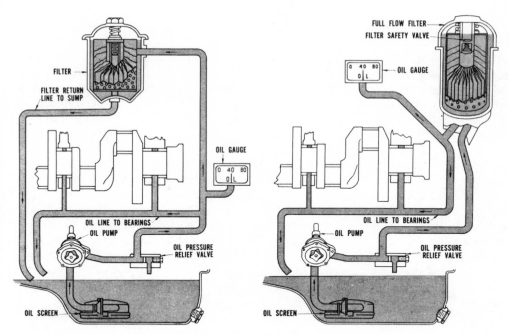

Fig. 12-23 Bypass and Full-Flow Systems
of Lubrication.

On some engines all the oil is passed through the oil filter as it comes from the pump. This is referred to as the full-flow system. Other engines have the bypass system in which only a portion of the oil is filtered at any given moment. The bypass system may filter 5% to 20% of the oil. However, over a period of time all the oil will be filtered.

The engine crankcase has a ventilation system designed to help keep the engine oil pure and to reduce air pollution. By-products of combustion may leak down into the crankcase, gasoline may leak into the crankcase under certain conditions, or moisture in the air may condense in the crankcase. These conditions can result in the formation of sludge and acids and the dilution of the oil. The ventilation system keeps these conditions to a minimum by providing a draft of clean air through the crankcase. Filtered air enters the crankcase, flows through it, and then is routed to the combustion chamber.

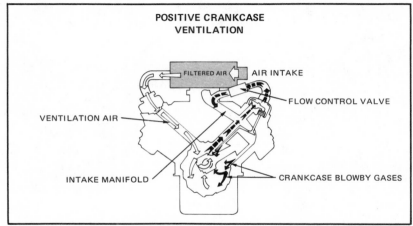

Fig. 12-24 Positive Crankcase Ventilation
Reduces Air Pollution.

DRIVE SYSTEM

The drive system transmits the power to the wheels which propel the automobile. The system contains four basic parts: (1) clutch, (2) transmission, (3) drive shaft and universal joints, and (4) differential. The clutch can engage or disengage the engine from the rest of the drive system. When the automobile is first started or standing still, the clutch is disengaged; when we want to move, the clutch is engaged.

The transmission enables us to change the speed of the wheels in relation to the speed of the engine; low speed, high torque for starting up; medium speed, medium torque for intermediate speed; and high speed, low torque for rapid travel. The transmission also provides a reverse gear for backing.

The drive shaft transmits the power to the differential. It is sometimes called the propeller shaft. The universal joints enable the axis of the transmission and the axis of the differential to be dissimilar. It has the effect of being a flexible shaft. The differential changes the direction of the drive, "carries it around the corner", since the drive shaft and rear axles are at right angles to each other. Also, the differential allows the wheels to rotate at different speeds for turning corners.

The single plate clutch is the most common type. It is mounted directly behind the engine. Its main parts are the flywheel, the friction disc, and the pressure plate. The polished flywheel, which is fastened to the end of the crankshaft, is always turning as long as the engine is running. The friction disc is faced with a material similar to brake lining material and mounted on the input side of the transmission. The end of the transmission shaft is splined so the disc can slide back and forth. The polished pressure plate is behind the disc and revolves with the flywheel. When the clutch pedal is out, strong springs push the pressure plate against the disc and up to the flywheel. The disc rotates with the flywheel; the clutch is said to be engaged. When the clutch pedal is pushed in, the spring pressure is released and the friction disc loses contact with the flywheel. The clutch is said to be disengaged.

The transmission enables the operator to change the relationship of the engine speed to the rotating speed of the wheels. This is done through the use of gears. By shifting the power to the various gears the proper torque and speed can be obtained. A large gear driving a small gear increases speed but reduces torque, while a small gear driving a large gear reduces speed but increases torque.

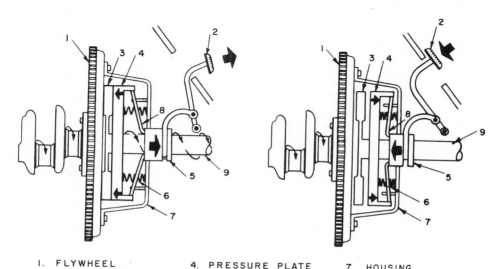

1. FLYWHEEL	4. PRESSURE PLATE	7. HOUSING
2. CLUTCH PEDAL	5. RELEASE BEARING	8. RELEASE LEVER
3. CLUTCH PLATE (DISC)	6. PRESSURE SPRING	9. CLUTCH GEAR BEARING RETAINER

Fig. 12-25 Clutch Operation.

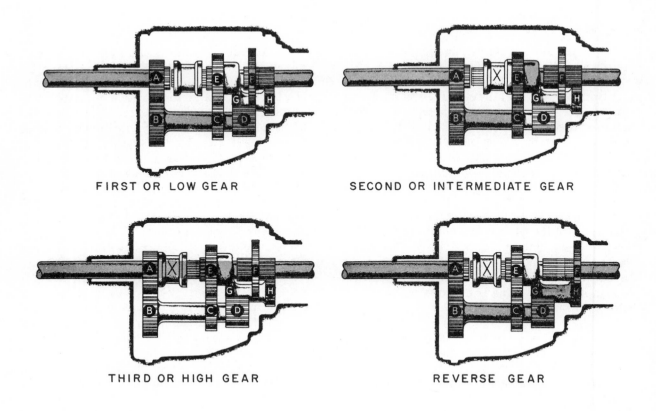

FIRST OR LOW GEAR   SECOND OR INTERMEDIATE GEAR

THIRD OR HIGH GEAR   REVERSE GEAR

Fig. 12-26   Conventional Constant Mesh Transmission.

The sliding gear transmission is the conventional type transmission providing low gear, intermediate gear, high gear, and reverse. In first or low gear the power follows these gears: (A) from the clutch, (B) on the counter shaft, (D) on the counter shaft, (F) on the transmission main shaft. This arrangement produces a gear reduction ratio of 3:1. Accompanied by a 4:1 reduction in the differential, the crankshaft turns twelve times for one turn of the wheels. Gears (E), (C), (G), and (H) are arranged to deliver no power in low gear.

In second gear or intermediate gear the power follows these gears: (A) from the clutch, (B) on the counter shaft, (C) on the counter shaft, (E) on the transmission main shaft. Gear (E) has been connected to the main transmission shaft by moving the positive clutch X to the right. Gear (F) is moved out of mesh. This arrangement produces a gear reduction ratio of 2:1. Again accompanied by a 4:1 reduction in the differential, the crankshaft turns eight times for one turn of the wheels.

In third or high gear the power from the clutch shaft goes directly to the main transmission shaft. The positive clutch X is moved to the left, engaging the clutch shaft and disengaging gear (E). Gear (F) is also disengaged. This direct connection provides a ratio of 1:1; however, it is accompanied by the 4:1 reduction in the differential. The crankshaft turns four times for one turn of the wheels.

In reverse the power follows these gears: (A) from the clutch, (B) on the counter shaft, (D) on the counter shaft, (G) on the reverse idler, (F) on the transmission main shaft. Gear (E) revolves but delivers no power. When this fifth gear is introduced the direction of the transmission shaft rotation is reversed. The gear ratios are the same as for low gear.

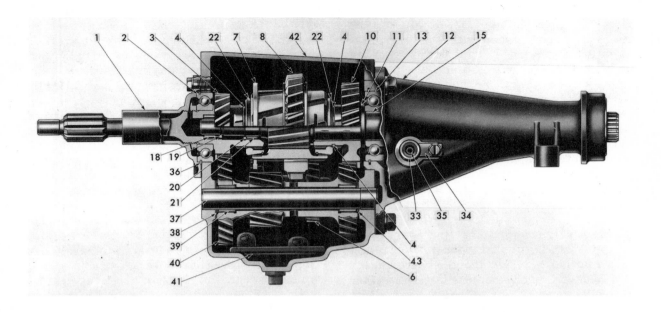

1. Clutch Gear Bearing Retainer	10. Second Speed Gear	20. Thrust Washer	37. Countershaft
2. Clutch Gear Bearing	11. Thrust Washer	21. Rear Pilot Bearing Rollers	38. Thrust Washer
3. Clutch Gear	12. Case Extension	22. Synchronizer Ring	39. Roller Bearing
4. Energizing Spring	13. Mainshaft Rear Bearing	33. Speedometer Shaft Fitting	40. Countergear
6. Reverse Idler Gear	15. Mainshaft	34. Lock Plate	41. Collector
7. Second and Third Speed Clutch	18. Front Pilot Bearing Rollers	35. Speedometer Driven Gear Shaft	42. Transmission Case
8. First and Reverse Sliding Gear	19. Thrust Washer	36. Snap Ring	43. Roller Thrust Washer

Fig. 12-27   Cross-section of Transmission.

The drive shaft and universal joints carry the power to the differential. The differential transmits the power to the axles and also enables the wheels to revolve at different speeds when a corner is turned.   The wheel on the inside of the turn needs to travel slower while the wheel on the outside of the turn needs to travel faster just as marching soldiers do in turning a corner.

The parts of the differential are (1) case, (2) drive pinion, (3) ring gear, (4) differential side bevel gears on the end of the axles, (5) pinion gears mounted in the differential case.

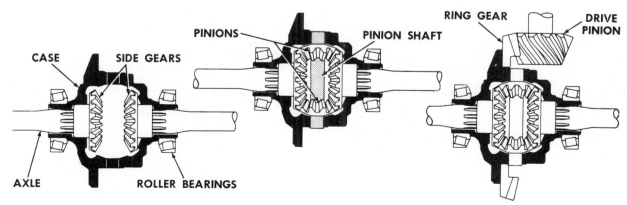

Fig. 12-28   Differential Changes the Direction of Power.

The drive pinion brings power into the differential and meshes with the ring gear. These two gears transfer the power from the drive shaft to the axles, which are at right angles to each other. The ring gear has about four times as many teeth as the drive pinion, the effect being a gear reduction ratio of 4:1. If the automobile always traveled in a straight line these would be the only parts necessary. However, to compensate for different rear wheel speeds upon turning, the differential side bevel gears and pinion gears are added. These gears are mounted in a case. The ring gear is also mounted on this case. When the auto travels straight ahead, the ring gear, case, and differential gears all move as a unit. But when a corner is turned the differential gears turn relative to each other, one faster, the other an equal amount slower. On slippery ice with one wheel spinning this wheel's differential side gear would be turning at twice ring gear speed while the other differential side gear is stationary.

Many automobiles are now equipped with automatic transmissions. The conventional clutch is eliminated, being replaced by a type of hydraulic drive. The simplest hydraulic drive is fluid coupling. It consists of two wheels fitted with blades, one connected to the crankshaft and called the pump or driver; one connected to the transmission shaft and called the turbine or driven member. The two are fitted close together but do not touch. They might be visualized as a doughnut sliced in two discs. These parts are in a sealed, oil-filled case. The driver blades throw the oil (exert pressure) against the driven blades when the engine is running. At idle speed, the pressure exerted is very small and the driven member does not move; the automobile does not move. The driven member is slipping 100 percent. However, as the engine speeds up the oil pressures against the driven blades increase and the driven member begins to revolve, slowly at first but faster and faster. Soon the driver and driven member are at almost the same speed, there is almost no slippage.

The automatic transmission works in conjunction with the hydraulic drive. The driver does no shifting; shifting is done automatically in the transmission, controlled by the speed of the auto and how far the accelerator is pushed in. The planetary gear is the heart of the auto-

LEGEND - Fig. 12-29

1. Transmission Housing
2. Converter Cover "O" Ring Seal
3. Turbine Assembly
4. Stator Assembly
5. Converter Housing & Pump Assembly
6. Converter Pump
7. Converter Pump Thrust Washer
8. Front Oil Pump Body Oil Seal
9. Front Oil Pump Body
10. Front Oil Pump Body "O" Ring Seal
11. Stator Support
12. Transmission Valve Body
13. Input Shaft Oil Seal Ring
14. Clutch Drum Oil Seal Rings
15. Clutch Relief Valve Ball
16. Low Brake Band
17. Clutch Drum
18. Clutch Piston Inner Seal
19. Clutch Hub
20. Clutch Hub Thrust Washer
21. Low Sun Gear & Clutch Flange Assembly
22. Parking Lock Gear
23. Planet Short Pinion
24. Planet Input Sun Gear
25. Planet Input Sun Gear Thrust Washer
26. Planet Carrier
27. Reverse Brake Band
28. Output Shaft
29. Transmission Case
30. Rear Oil Pump Gasket
31. Rear Oil Pump Cover to Body Attaching Screw

32. Rear Oil Pump Cover
33. Rear Oil Pump Body
34. Rear Bearing Locating Front Snap Ring
35. Transmission Rear Bearing Assembly
36. Transmission Rear Bearing Retainer
37. Rear Bearing Locating Rear Snap Ring
38. Transmission Extension "O" Ring Seal
39. Transmission Extension
40. Speedometer Drive Gear
41. Extension Rear Oil Seal
42. Extension Bushing
43. Speedometer Driven Gear
44. Transmission Rear Bearing Retainer Screw
45. Transmission Rear Bearing Retainer Screw Lockwasher
46. Rear Oil Pump Drive Gear Drive Pin
47. Rear Oil Pump Assembly Attaching Screw
48. Rear Oil Pump Drive Gear
49. Rear Oil Pump Driven Gear
50. Governor Drive Gear
51. Governor Driven Gear
52. Transmission Case Bushing
53. Reverse Drum Thrust Washer
54. Planet Long Pinion
55. Reverse Band Lever & Link Assembly
56. Low Sun Gear Thrust Washer
57. Planet Pinion Shaft Lock Plate
58. Reverse Drum & Ring Gear
59. Clutch Flange Retainer
60. Clutch Flange Retainer Ring

61. Clutch Spring Seat
62. Clutch Spring Snap Ring
63. Clutch Spring
64. Clutch Drive Plates
65. Clutch Driven Plates
66. Clutch Piston
67. Clutch Piston Outer Seal
68. Clutch Drum Thrust Washer (Selective)
69. Manual Valve
70. Converter Housing Dowel Pin
71. Converter Housing-to-Case Gasket
72. Front Oil Pump Drive Gear
73. Front Oil Pump Driven Gear
74. Sump Baffle
75. Oil Pickup and Suction Screen
76. Access Hole Plug
77. Converter Pump Housing Bolt
78. Converter Pump Housing Nut
79. Stator Retaining Rings
80. Stator Thrust Washers
81. Over-Run Cam Roller
82. Stator Race
83. Converter Cover Hub Bushing
84. Input Shaft
85. Turbine Thrust Washer
86. Over-Run Cam Roller Spring
87. Over-Run Cam
88. Converter Cover Assembly
89. Flywheel to Transmission Anchor Nut Assy.

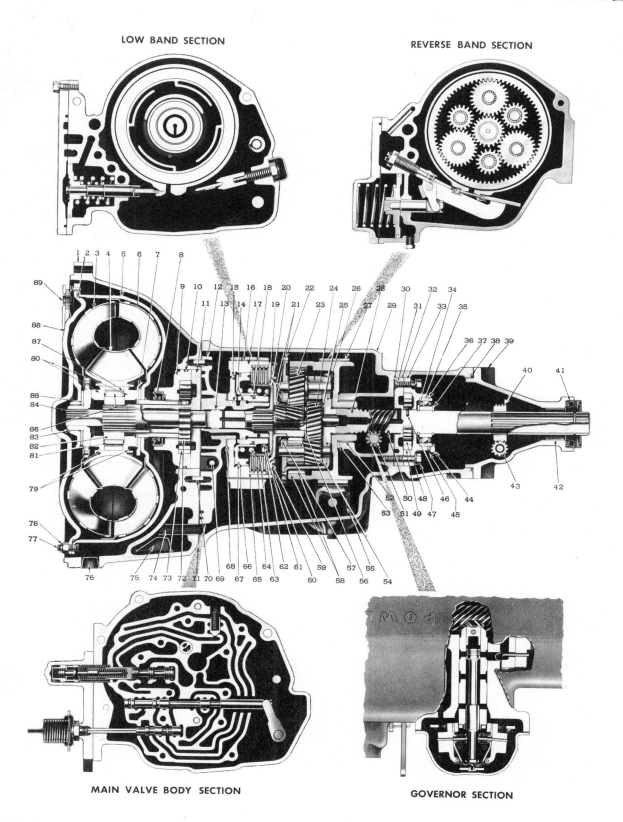

LOW BAND SECTION

REVERSE BAND SECTION

MAIN VALVE BODY SECTION

GOVERNOR SECTION

Fig. 12-29 Cross-section of Torque Converter.

matic transmission. Often there are three planetary gear sets in the transmission, two for four forward speeds and one for reverse. The planetary gears are always in mesh and do not have to be shifted away.

The planetary gear is made of three main parts: (1) the center or sun gear, (2) planet pinion gears, and (3) internal or annulus gear. The gears are always in mesh. The three planet pinion gears are held 120° apart by the planet carrier.

The correct gear reduction is attained by holding one of the gear elements, allowing the power to be transmitted in either direction between the other two elements. To illustrate, if the internal gear is held stationary and power is applied to the center gear, the planet pinion gears revolve inside the internal gear. Power is taken from the planet pinion's planet carrier at a certain gear reduction ratio.

Also, if the center gear is held stationary and power is applied to the internal gear, the planet carrier will revolve around the center gear. Gear reduction is again attained, but at a different ratio.

To obtain reverse rotation in the transmission, the planet carrier is held stationary and power is applied to the center gear. The planet pinion gears spin on their bearings acting

as idlers and transfer the power to the internal gear. Power is taken from the internal gear which is rotating in reverse direction.

If none of the planetary gear elements are held stationary no power is transmitted, the transmission is in neutral. The planetary gear transmission is operated by oil pressure which tightens bands and clutches onto the gears at the proper time.

The hydraulic torque converter is another type of automatic transmission. It is very similar to fluid coupling but it has the ability to multiply the torque or twisting force produced. The torque converter also has a driver and driven member, pump and turbine, but in addition it has a set of stationary blades that receive the oil from the turbine and change its direction of flow. This change of direction helps the pump to pump harder when the oil re-enters, thereby multiplying the torque up to two and one-third times under certain conditions. Once the automobile has attained speed, the torque converter acts as a simple fluid coupling. Planetary gears are usually used with the torque converter to obtain reverse or to obtain a special low gear for heavy pulling.

NOTE: For the information on pollution controls and the automobile engine, refer to Unit 25, Power Technology and the Environment, pages 276-280.

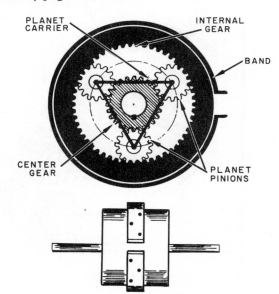

Fig. 12-30   Planetary Gear System.

SAFETY IN AUTOMOTIVE MAINTENANCE AND REPAIR

The repair and maintenance of auto engines and the associated parts of the automobile present conditions that can be a safety hazard if safety rules are not followed in an intelligent manner. Accidents can quickly convert a truly satisfying activity into a grave personal tragedy.

Before discussing safety rules as such, every student mechanic or professional should be fully aware of the responsibility involved with auto repair work. There is absolutely no room for halfway or makeshift work. Errors can place a life in jeopardy and therefore they cannot be tolerated.

The mechanic's attitude must be one that demands perfection —

  √ no mistakes

  √ every nut and bolt properly tightened

  √ every part in its proper place

  √ every part functioning correctly

Burns

1. Take extreme care in draining radiators, being careful to protect hands and face when removing the cap from a boiling radiator.

2. For protection against battery acid burns, wash the hands immediately after having come in contact with battery acid.

3. Avoid coming in contact with exhaust pipes, manifolds, and other parts of an engine which is hot.

Falling Objects

1. Be sure that chain hoists are fastened securely and that chain slings are not overloaded.

2. Do not depend on wheel jacks or hoists alone. Always use strong wooden blocks or a substantial steel or wooden horse made for the purpose.

3. Allow no one under the car while it is being lifted.

4. When two wheels are off of the floor, keep the other wheels blocked to prevent the car from rolling.

5. When working under a car, keep feet and legs clear of passageways.

Fire

1. Splashing gasoline may start a fire on a hot motor. Never run a car above idle speed when the float bowl cover is removed.

3. Store gasoline in an approved safety can.

4. Be sure that all liquid containers in the shop are labeled properly.

5. Do not use gasoline to clean tools, machine parts, or clothing. Use kerosene for cleaning oil and grease from metal parts and tools.

6. Never wash hands or arms in gasoline.

7. If clothing has been soaked with gasoline, remove it as it may cause skin trouble.

8. Put gasoline soaked rags in safe covered metal waste cans only.

9. When pouring gas from one container to another, keep containers in contact to bleed off static.

10. Locate fire extinguishers where handy. Use only foam or carbon dioxide on gas or oil fires.

11. Engines operated indoors should be provided with flexible exhaust tubes attached to the tailpipe to carry dangerous carbon monoxide gases outdoors.

Shocks

1. When working on a car in which the motor need not be in operation, disconnect the battery and insulate the connections.

2. Check the extension cords to see that they are well insulated.

3. Check portable electric machines to see that they are grounded properly.

125

Lifting

1. Lift with the legs, not the back. Use hoists wherever possible. Get help on heavy lifting.

Cuts

1. Be careful when removing bulbs and lenses from the headlights. Use proper tools to prevent cuts in case of breakage.

2. In closing windows, never hit the panes; use the handles.

3. Use great care in replacing broken glass in the car.

Cranking

1. Use a safety grip (thumb not around handle) when cranking a car.

2. See that the transmission is in neutral and the spark retarded. Stand a little back from the crank. Do not straighten the arm at the elbow.

3. Avoid spinning the engine.

Moving Parts

1. Do as much work as possible on the engine when it is stopped.

2. Do not lubricate the engine while it is in motion.

3. Do not use a wiping cloth on moving parts of the engine.

4. Do not get the hands in door jambs when cleaning windshields or when performing other similar tasks.

Hoists

1. Always use a safety device on a hoist designed to keep it from falling.

2. Make sure the car is out of gear with the motor turned off but the brakes set.

3. Check the wheel blocks before lifting.

4. Keep from under the hoist while the car is being raised or lowered.

Lubricating Rack

1. Have frequent inspections of the air hose. Replace worn hose because it may burst under pressure.

2. Allow no fooling with air hoses or with high pressure lubrication equipment. Air hoses are dangerous when used in horseplay or for dusting off clothing or hair.

3. Do not allow students to point the gun at any person even though it is grease clogged. High pressure lubrication equipment can cause injury to face or body.

Tire Inflation

1. Check pressure gages frequently to see that they are working properly.

2. Instruct students never to face a tire while it is being inflated. Excessive pressure may cause the tire to explode.

Unsafe Tools

1. Have the person at the tool crib inspect all tools as they are issued and as they are returned. Unsafe tools should not be allowed in the shop.

2. Use goggles on all jobs that have an eye hazard, such as grinding or chipping.

3. Have students report any unsafe tool or equipment to the instructor at once.

Horseplay

1. Do not allow running, scuffling, or throwing tools and materials in the shops. Thoughtless acts such as these can result in serious injury.

Some General Rules for Auto Shops

1. Have keys removed from all cars in the school shop and turned over to instructor.

2. Allow only the instructor to start the motor in any car in the shop.

3. Allow only one student to work on the inside of a car at a time.

GENERAL STUDY QUESTIONS

1. Explain the job of the rocker arms.

2. Explain why correct engine timing is essential.

3. The fuel pump is located between what two parts of the fuel system? Explain the operation of the fuel pump.

4. Why are the idle system, power system, and acceleration system necessary on a carburetor?

5. What is the key part of the automatic choke?

6. What parts make up the exhaust system?

7. List the main parts of the ignition system.

8. Explain the difference between full-flow and bypass lubrication.

9. Why is a ventilation system necessary?

10. Why is the differential necessary?

11. On many automatic transmissions what replaces the conventional clutch?

12. What gears are the "heart" of the automatic transmission? List their parts.

13. What is a hydraulic torque converter?

CLASS DISCUSSION TOPICS

● Discuss why the automobile engine may need more refined systems than a small engine: heavier loads, different operating speeds, greater need for economy, etc.

● Discuss the basic parts of the engine and explain the function of each.

● Discuss the essentials of the fuel system and why it is necessary to provide several filtering screens in the system.

● Discuss the operation of the fuel pump.

● Discuss the reasons for having an idle, main, power, and accelerating system in the carburetor.

● Discuss the operation of the carburetor systems, pointing out carburetor parts.

● Discuss the engine's lubrication system; its main and auxiliary functions, how the oil moves through the engine, the action of the oil pump, and the construction of the oil filter.

● Discuss the operation of the single plate clutch.

● Discuss the operation of the sliding gear transmission.

● Discuss the operation of the hydraulic drive.

● Discuss the operation of the automatic transmission.

● Discuss the operation of the torque converter.

DIESEL ENGINES

For centuries, the natives of the South Pacific have employed an ingenious method for lighting fires. Using a bamboo cylinder in which dry tinder is placed, a close-fitting plug is driven sharply into the open end. The air entrapped in the cylinder is compressed and heated to the point where the tinder ignites. This spontaneous combustion from the heat of compression is the principle of compression ignition, the basis on which the diesel engine operates.

The characteristic, then, of the diesel engine which distinguishes it from other internal combustion engines is the absence of any external ignition device. The gasoline engine uses an outside spark source to ignite the fuel mixture. This type of engine is thus known as a spark-ignition engine. The diesel, however, uses only the terrific heat generated by the compression of air in the cylinder to ignite the fuel. Thus, the diesel is a compression-ignition engine.

The inventor of the diesel engine, Dr. Rudolph Diesel, was born in 1858 of German parents. He became interested in engines during his college years. He was graduated as an engineer when he was 21 years old and he immediately set out to build an engine that would be more efficient than the low efficiency steam engines of the time. His first engine was patented in 1892. But when built, it exploded as he tried to start it. However, he escaped serious injury in the unfortunate explosion. The tragedy did not cause him to give up; he was further convinced of the correctness of his basic ideas. A few years later, 1897, Dr. Diesel did produce a successful engine, a single-cylinder engine capable of developing 25 hp.

The first commercial diesel engine was put into service a year later in St. Louis, Mo. It was a two-cylinder, 60-horsepower engine. The use of diesel engines spread rapidly and, in a few years, thousands of engines were in use.

The early engines were used mostly to replace steam engines in large installations such as stationary power plants or large boats. The first engines were heavy, weighing as much as 250 pounds per horsepower. They could not compete with the gasoline engine used on land vehicles where weight was a more important factor. Today, diesel engines are much lighter in weight and, of course, are commonly found on all types of trucks, heavy road machinery, and railroad locomotives. Some of today's engines weigh as little as 7 pounds per horsepower.

EXTERNAL COMBUSTION

INTERNAL COMBUSTION

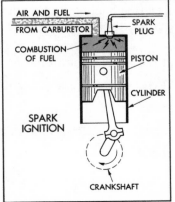
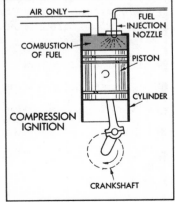

Steam Engine   Gasoline Engine   Diesel Engine

Fig. 13-1 Comparison of Combustion Systems.

A major difference between diesel and gasoline engines is the compression ratio for which each engine is designed. Compression ratio is defined as the ratio of the cylinder and clearance volume with the piston on bottom dead center to the clearance volume alone with the piston on top dead center.

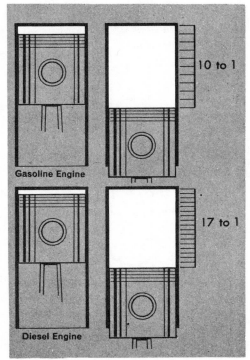

Fig. 13-2 Compression Ratios of Gasoline and Diesel Engines.

Diesel engines are all "high compression" engines, usually in the range of 17 to 1 versus 10 to 1 for gasoline engines used on automobiles. Compression ratios in the nature of 17 to 1 are not possible in a carburetor-fed gasoline engine even using the best anti-knock gasoline. A mixture of air and gasoline would ignite long before the piston reached top dead center because of the intense heat caused by the high compression.

The high compression ratio gives the diesel its efficiency advantage. We know that when the stored energy of a fuel is released or burned, all of the energy does not do useful work. Much of the heat energy is lost in waste exhaust gases, lost to friction, or lost in heating up the engine. Of the common engine types, the

diesel engine is the most efficient: more of the potential energy in the fuel is put to work. The simple steam engine has an efficiency of 6% to 8%. The steam turbine and condensing engine has an efficiency of 16% to 33%. The gasoline engine has an efficiency of 25% to 32%. The diesel engine has an efficiency of 32% to 38%. The more efficient the engine, the less fuel it will have to consume to produce comparable power.

The construction of gasoline and diesel engines is similar in many respects; crankcase, crankshaft, cylinder, piston, etc. The main differences between the engines lie in these four points:

1. How the air enters the cylinder.

2. How the fuel enters the cylinder.

3. How the fuel is ignited.

4. How the exhaust gases leave the cylinder.

The diesel can be either a four-stroke or a two-stroke cycle engine. Therefore, the answers to the "differences" listed above depend on which engine is discussed.

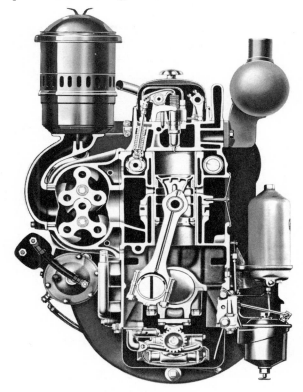

Fig. 13-3 Cross-section (End View) of a Diesel Engine.

In a two-stroke cycle diesel the air enters the cylinder by being "blown" in through intake ports which are located near the bottom of the cylinder. The fuel enters the cylinder at the top of the compression stroke through a fuel injector. The fuel is ignited by the high temperature of the compressed air and not by a spark plug. Upon completion of the power stroke, the exhaust gases leave when exhaust valves are opened and, at about the same time, new air is forced into the cylinder scavenging out the last bit of exhaust.

In a four-stroke cycle diesel the air enters the cylinder the same as it would in a gasoline engine, rushing to the low pressure area in the cylinder. However, it does not pass through a carburetor. The fuel enters the cylinder when the piston reaches the top of the compression stroke. It enters through a fuel injector. The fuel is ignited by the high temperature of the compressed air and not by a spark plug. The exhaust gases leave as they would in a gasoline engine, being pushed out by the upward motion of the piston on the exhaust stroke.

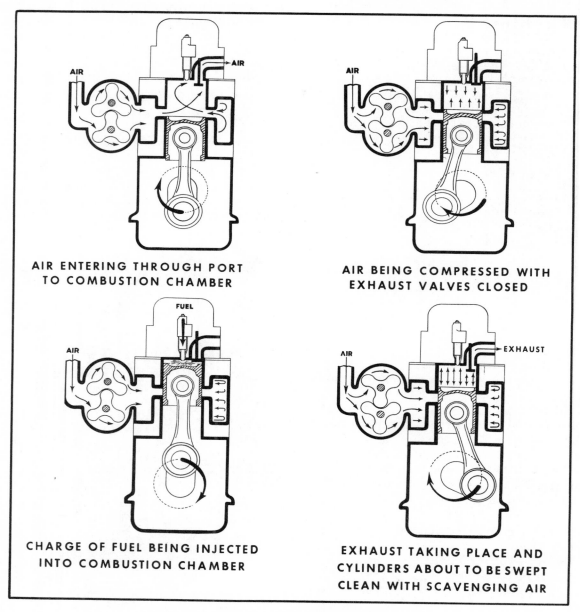

AIR ENTERING THROUGH PORT
TO COMBUSTION CHAMBER

AIR BEING COMPRESSED WITH
EXHAUST VALVES CLOSED

CHARGE OF FUEL BEING INJECTED
INTO COMBUSTION CHAMBER

EXHAUST TAKING PLACE AND
CYLINDERS ABOUT TO BE SWEPT
CLEAN WITH SCAVENGING AIR

Fig. 13-4 Two-Stroke Cycle Diesel Engine.

FUEL INJECTION

The fuel injector eliminates the need for a carburetor on the diesel. With the air tightly compressed and at a high temperature, the fuel injector sprays diesel oil into the cylinder. The oil enters as a mist or fog and begins burning immediately upon contact with the super-heated air. No spark ignition system is necessary since the fuel begins burning on its own when it contacts the air.

The fuel injector parts are machined to extremely exacting tolerances. The unit is a precision machine in its own right. For example, on a unit injector the injection pump piston is fitted to the pump cylinder to an accuracy of 30 to 60 millionths of an inch. Such accuracy is necessary since the fuel must be forced through the injector nozzle in a split

second and at a tremendous pressure. The fuel may be forced through the several needle-point sized holes at the nozzle tip with a pressure of 3,000 to 30,000 pounds per square inch. Of course, there cannot be leakage between the plunger and cylinder, and the injector must meter out the exact amount of fuel called for. Each cylinder of the engine is fitted with its own fuel injector.

To vary the engine speed, the amount of fuel sprayed from the injector is changed. This is another factor in the complexity of the fuel injector. Just the right amount of fuel is metered out for the desired speed. Fuel only is changed: the amount of air taken into the cylinder and compressed remains the same throughout the speed range.

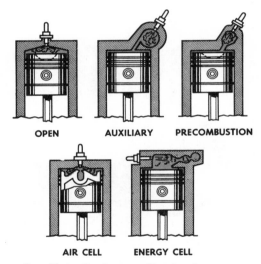

Fig. 13-6 Combustion Chamber Designs.

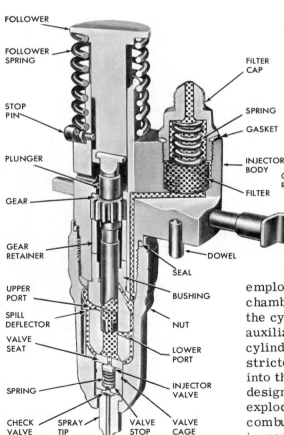

Fig. 13-5 Fuel Injector Assembly.

Various types of combustion chamber designs are employed to improve the combustion in the main cylinder chamber. In the open design, fuel is injected directly into the cylinder between the piston and the cylinder head. The auxiliary design includes a small chamber to the main cylinder. Fuel is injected into the chamber and the restricted passage increases the turbulence of the discharge into the main chamber. Similarly, in the precombustion design, burning begins in the precombustion chamber and explodes violently into the main cylinder for complete combustion. The air cell design, again, is a means of increasing the force of the combustion. The energy cell, however, with its chamber opposite from the fuel injector, tends to produce a smooth buildup of pressure in the main chamber rather than a violent burst.

AIR BLOWER

The four-stroke cycle diesel engine gets its air supply much the same as a gasoline engine does: the air is drawn in as the piston goes down the cylinder. However, on a two-stroke cycle engine, the same principle cannot be used since exhaust and intake must take place at the same time. As the exhaust gases leave through valves located at the top of the cylinder, air is pumped in through ports located near the bottom of the cylinder.

There are several air pumps used but one of the most common types is known as a Roots blower. It consists, basically, of two three-lobed, intermeshing spiral gears which trap the air and force it to the opposite side. It turns constantly and pumps air under pressure to the cylinder. Its action might be compared to that of a revolving door bringing people in from the street.

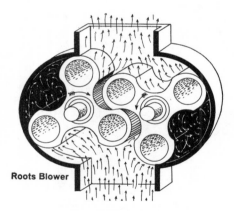

Fig. 13-7   The Roots Blower.

Some diesels use a second blower, a turbocharger, in addition to the Roots blower. The turbocharger is a centrifugal pump driven by exhaust gases or by a turbine which is located in the exhaust section of the engine. The turbocharger reduces the demands placed upon the Roots blower: it may be now driven slower or be a smaller sized blower.

Diesel engines burn their own special type of fuel. It is distilled petroleum fuel oil and is produced in much the same manner as gasoline. It is common for people to think of diesel fuel

as "cheap fuel", "low quality." Nothing could be further from the truth. The quality of the fuel is high and it is produced to exacting standards. It must burn leaving little or no deposit, be very clean, and ignite properly. Surprisingly, the price of diesel fuel and gasoline is almost the same at the refinery. The cost difference is mostly due to greater taxation on gasoline. The economy of operating a diesel does not come from burning cheaper fuel, rather it comes from burning the fuel in a very efficient engine.

It will be helpful to "walk through" one cycle of a two-stroke cycle diesel engine. At the top of the compression stroke, the air is tightly compressed and is at a temperature of 1000° F. Now the fuel injector sprays an oil fog into the air. The oil begins burning immediately and the extreme pressures caused by combustion push the piston rapidly back down the cylinder. When the piston nears the bottom

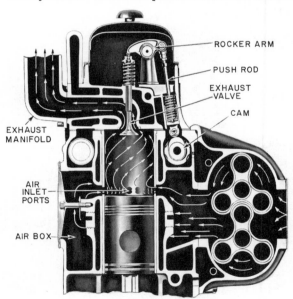

ROCKER ARM

PUSH ROD

EXHAUST
VALVE

CAM

EXHAUST
MANIFOLD

AIR
INLET
PORTS

AIR BOX

Fig. 13-8   Intake and Exhaust System
Two-Cycle Stroke Diesel.

of its stroke, intake ports are uncovered and air is blown into the cylinder. Also, at this time, the exhaust valves located at the top of the cylinder open, allowing exhaust gases to leave. The last amount of exhaust is scavenged from the cylinder by the incoming air. Exhaust valves now close, the piston moves upward sealing off the intake ports, and the air charge is trapped and compressed.

The world of the diesel is constantly expanding. One of the most notable applications of the diesel engine is the diesel locomotive. This use has made the steam locomotive nearly obsolete. Diesels can effect savings in operation and maintenance. Diesel engines for boats, trucks, road machinery, busses, off-shore oil drilling rigs, and transportable electric power units are other applications where the diesel is very effective.

GENERAL STUDY QUESTIONS

1. What were early diesel engines used for?

2. Why can the diesel produce more power with a given amount of fuel energy than other engines?

3. What does the diesel's high compression ratio have to do with its success?

4. Why doesn't the diesel use a carburetor?

5. What temperature does the air reach under compression?

6. What causes the fuel to ignite in the combustion chamber?

7. How does the fuel get into the cylinder?

8. What is the purpose of an air blower on a diesel engine?

9. When engine speed is changed what is varied: fuel, air or both fuel and air?

10. Explain why diesel fuel is not cheap or low-grade fuel.

CLASS DISCUSSION TOPICS

- Discuss the modern applications of the diesel engine.

- Discuss molecular motion on the compression stroke and explain how heat is created.

- Discuss the advantage of a high compression engine and why it is not easy to build the feature into smaller engines.

- Discuss efficiency in engines and explain how this affects overall engine performance.

- Discuss the advantage of a two-stroke cycle diesel over a four-stroke cycle diesel.

- Discuss the possibilities of fuel injection for gasoline engines.

- Discuss the possibilities of blowers or superchargers for gasoline engines.

Unit 14

JET ENGINES

Most of us think of jet propulsion as a "new idea", one developed during this century. True, the most notable application of jet propulsion, the jet aircraft, has come about during the past few decades, but the theory and limited use of jet propulsion stretches back for centuries. Hero of Alexandria invented the Aeolipile one hundred years before the birth of Christ. His machine illustrates the principles of jet propulsion.

Hero's Aeolipile was basically a covered vessel containing water. Two tubes mounted on top of the vessel acted as an axis for a hollow sphere. On the sphere were two nozzles with their openings 90° to the axis of rotation and opposing each other. A fire built below the vessel boiled the water and the steam traveled up the tubes, into the sphere, and out the nozzles. The pressure of the escaping steam caused the sphere to rotate rapidly. The idea was regarded as a novelty, a toy, not to be developed into a useful machine for centuries.

Newton's third law of motion explains why a jet engine works and why Hero's Aeolipile spins. This law states: For every acting force there is an equal and opposite reacting force. In the Aeolipile the acting force of the escaping steam created as the water within the container boils causes an equal and opposite reacting force in the spinning of the sphere.

A toy balloon can also illustrate the jet principle. Blown up and held closed, the balloon has no motion; pressure inside the balloon is equal in all directions. However, when the balloon is released, its elasticity causes the air to be forced out rapidly. This action is accompanied by an equal but opposite reaction that can be seen as the balloon is propelled forward.

In a jet engine tremendous pressures build up inside the engine during combustion. These hot gases exert their pressure in all directions but the escape route to relieve the pressure is out the rear of the engine. In simple terms, the action of the gases rushing rearward produce the reaction of moving the engine forward.

Fig. 14-1   Hero's Aeolipile Illustrates the Principles of Jet Propulsion.

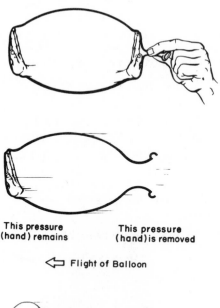

This pressure (hand) remains   This pressure (hand) is removed

⟸ Flight of Balloon

Fig. 14-2   A Toy Balloon Also Illustrates Jet Propulsion Principles.

Fig. 14-3   A Modern Jet Engine, General Electric's J79-11A.

The rapid ejection of gases produces <u>thrust</u>, the driving force that pushes the engine (aircraft) forward. The jet is <u>not</u> pushing against the atmosphere as is often thought. The thrust comes from inside the engine itself when the gases accelerate to the opening, moving the engine forward. Jet propulsion power plants are, therefore, broadly classed as reaction engines. This type of engine produces a forward thrust by the forceful ejection of matter from within itself.

Jet engine power is measured in pounds of thrust produced. For example, a certain engine is described as capable of 10,000 pounds of thrust.

In this Unit the main types of jet engines will be discussed. Their basic construction and operating principles will be examined. In the first few paragraphs the principles of jet propulsion were briefly considered. In further discussion we will see how the principles are applied. The four types of jet engines studied are: (1) Ramjet, (2) Pulsejet, (3) Turbojet, and (4) Turboprop.

It might be noted that virtually all applications of jet engines are in the aviation field; aircraft, drones, etc. Jets are used, to a degree, in large power speed boats and experimentally in motor vehicles, stationary power plants, ships, and locomotives. Speed is the answer to the jet's popularity. We can travel faster than we ever dreamed possible with the aid of the jet engine. Today, travel to any corner of the earth is a matter of hours.

AIR SUPPLY - ROCKET OR JET

All jets burn oxygen from the earth's atmosphere with their fuel. We say the jet is an <u>air-dependent</u> engine; it cannot leave the atmosphere and still operate. A rocket engine is an <u>air-independent</u> engine; the relatively thin layer of the earth's atmosphere does not limit its range. The rocket engine carries its own oxidizer for its fuel. This is really the main distinction between jet and rocket engines since they both operate on the same physical principles, <u>action</u> and <u>reaction</u>.

The air mass around the earth is dense close to the earth and thins out as the altitude increases. Eventually, the earth's layer of atmosphere becomes nonexistent. Anyone can see how the atmosphere thins out by driving into the mountains. At high altitudes there is less oxygen and a "flatlander" must breathe more deeply. The thinning atmosphere greatly limits the altitude of the propeller-driven aircraft; its propellers need to bite into atmosphere, the thicker the better. The jet eliminates this propeller efficiency as an operating factor. The jet aircraft performs better at high altitudes and air speeds than it does at low altitude and speed.

RAMJET (Athodyd)

The ramjet is the simplest of the jet engines. It has no moving parts. Technically, it is an aero-thermodynamic duct. It is a cylinder, open at both ends, having a fuel injection system inside.

Air enters the front end of the cylinder and burns with the fuel being sprayed into the combustion section. The action of the high pressure exhaust gases leaving the cylinder produces the thrust that pushes the jet forward.

As the name "ram" implies, air is pushed into the engine when the jet speeds through the air. Air is literally "rammed" into the open cylinder.

It is impossible for this jet to take off on its own power. It must be accelerated to high speed by an auxiliary power before the ram effect can take over. If the pressure of the incoming air were too small due to slow speed, the burning fuel and air would tend to blow out both ends of the jet. The pressure of the ram air must be enough to keep the combustion gases traveling to the rear only.

The diffuser section (foremost section) is designed to get the correct ram effect. Its job is to decrease the velocity and increase the pressure of the incoming air. To do this, it is convergent: that is, it has a small entrance diameter sloping back to a larger diameter. This design increases the air pressure to the point where the combustion gases will not overcome the ram air pressure.

The fuel and air burn in the combustion section. Fuel injectors spray fuel into the incoming air and the operation is continuous. Initial starting is by a spark plug but once the engine starts there is no interruption of power. The incoming stream of fuel and air is ignited by the hot combustion gases. A flame holder toward the rear of the jet retards the gases so they will not be swept too far to the rear of the engine.

The ramjet is an excellent high altitude power plant. Its performance at high speed and high altitude is the best of the four jet engines. Its characteristics make it useful as an auxiliary power source mounted on the wing tips of high-altitude jet aircraft. Also, it is suitable for jet helicopters, the ramjet being mounted on the rotor blade tips. The ramjet has a high power-to-weight ratio; however, it does consume fuel at a rapid rate.

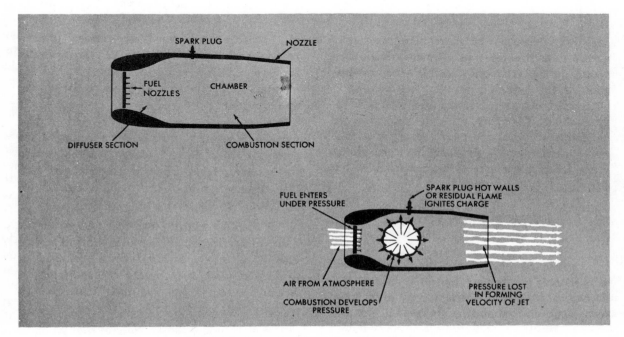

Fig. 14-4 Operating Principles of the Ramjet.

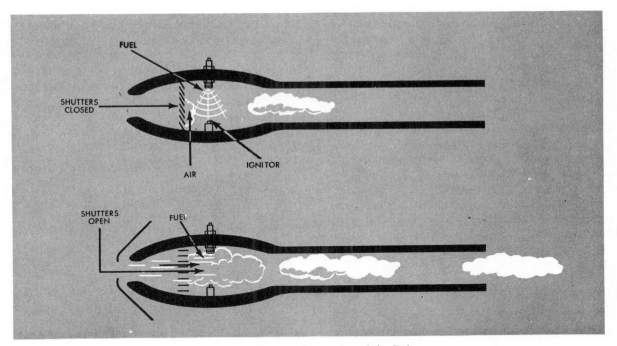

Fig. 14-5 Operating Principles of the Pulsejet.

PULSEJET

The pulsejet was developed into an operational engine by the Germans during the 1930's. Their application of the engine was the V-1 or Buzz Bomb which was used primarily against England during World War II. The Buzz Bomb was unmanned and, once aloft, it maintained an altitude of about 10,000 feet with a preset gyroscopic system and barometric unit. When the bomb reached target area, a measuring device cut off the fuel and the bomb went into a spiraling dive to the earth. Speed was 400 m.p.h. and many of the bombs were shot down by anti-aircraft crews and fighter pilots. Although the bomb had many drawbacks, it was effective for bombing a large metropolitan area such as London.

Also, during World War II a pulsejet missile was developed in the United States. It was known as the JB-2. However, the war ended before it could be put into use.

Today, the pulsejet is used primarily for drones and test vehicles. The pulsejet is quite simple and economical to construct. Most pulsejets need to be boosted into operation. Once in operation their speeds are moderate, being limited by the design of the engine.

As this name implies the pulsejet has a pulsating power: that is, the combustion is broken and not continuous. The pulses of power are, of course, very close together. For example, there may be 200 combustions per second. Other jets have continuous burning of the fuel.

The main parts of the pulsejet are: (1) the diffuser, (2) the grill assembly, (3) the combustion chamber, and (4) the tailpipe.

The ram air comes into the engine through the diffuser section. This high velocity air travels into an expanding cross-sectional area, the result being to decrease the velocity and to increase the air pressure.

The grill assembly is made up of a number of flapper units (shutters). The flapper unit has spring steel strips (similar to a reed valve on two-stroke cycle engines) that can spring shut upon combustion. Ram air pressure causes the flapper unit to open, but when combustion takes place, the combustion pressures overcome the ram air pressure and the flappers spring shut. These flapper units are the key to the pulsating operation of the engine.

The grill assembly also contains the fuel injectors and ventures, both located just behind the flapper units. The ventures insure proper mixture as fuel is sprayed into the airstream.

The combustion chamber area provides the area for the burning gases to expand. The initial combustion is begun by a spark plug but once the engine is operating the spark ignition is shut off. Flame from the tailpipe is drawn back up to ignite each succeeding fuel charge.

The tailpipe section with its smaller cross-section accelerates the exhaust gases. The length of the tailpipe determines the number of cycles per second; the shorter the tailpipe, the greater the frequency of cycles.

TURBOJET

The principle of the turbojet engine goes back several centuries but really active development of the engine began during World War II. Both England and Germany developed high-speed turbojet-powered fighters. First German work began in 1936 and the world's first turbojet-powered aircraft was flown in 1939. The airplane was the HE-178. A later German aircraft, the ME-262, was capable of speeds up to 500 m.p.h. Over 1600 of this class aircraft were produced near the end of World War II.

Jet engine work in England ran much the same course as that in Germany. In 1936 design and development began and in 1937 the engine was ground tested. Air Commodore Frank G. Whittle was the guiding force behind the English effort. In 1941 England's first jet-powered aircraft was flown; its name, the Gloucester Pioneer. Its successor, the Gloucester Meteor, was the only allied jet aircraft to become operational during the war.

The United States entered the jet aircraft field in 1941 when the General Electric Company was awarded the contract to improve and produce the English Whittle engine. The jet fighter XP-80 was the first operational U.S. fighter. It was built by Lockheed and powered by a General Electric GE-J-33 engine. This aircraft, later designated the F-80 Shooting Star, was capable of speeds of over 600 m.p.h.

In a short time jet fighters replaced propeller-driven fighters among all major nations. The constant improvement of turbojet engines soon led the way to their use on fighter bombers. Now bombers such as the B-52 are jet-powered. Perfected aerial refueling techniques along with engine improvements have made the intercontinental bomber a mainstay in our nation's defense structure.

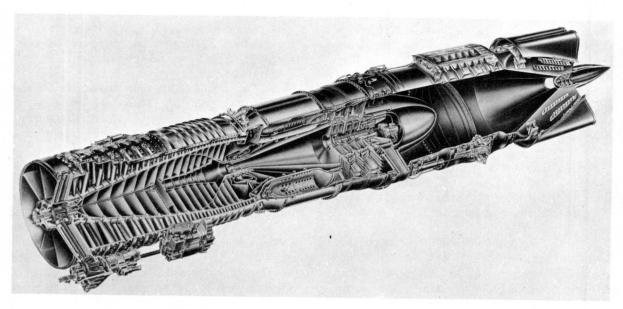

Fig. 14-6   Cutaway of General Electric Turbojet Engine CJ-805-3.

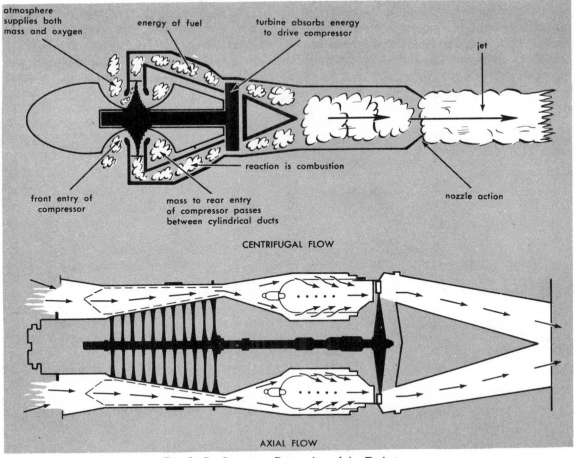

Fig. 14-7 Operating Principles of the Turbojet.

Of the jet engine types the turbojet is the most complicated, the reason being the inclusion of a gas turbine as a major engine component. With the turbine included the engine can "take off" on its own power: auxiliary boosters are not necessary as they are in the ram and pulsejets.

The air entering a turbojet goes through the following sections of the engine: Air Inlet, Compressor, Combustion Chambers, Turbine, Exhaust Cone, Tailpipe, and Jet Nozzle.

The compressor increases the pressure of the incoming air and delivers that air to the combustion chambers. As the amount of air delivered to the combustion chambers increases, the thrust produced increases. The amount of air depends on three factors: (1) the speed at which the compressor is turning, (2) the aircraft speed, and (3) the density of the air. Increasing any or all of these will put a greater mass of air through the engine.

The compressed air supports the combustion of fuel that is being sprayed into the combustion section. The expanding gases rush rearward to the turbine blades. As the gases travel through the turbine section a considerable amount of their energy is spent in turning the turbine. The turbine's motion is applied directly to turn the compressor.

About two-thirds of the engine's power works to rotate the compressor. The remaining one-third of the power goes out the exhaust cone, tailpipe, and jet nozzle. This one-third produces the thrust that pushes the aircraft forward.

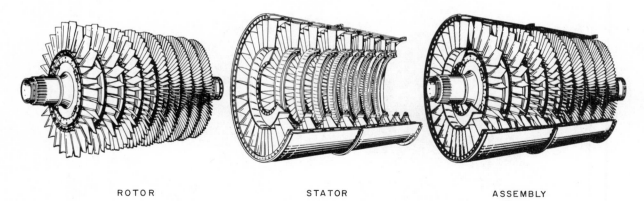

ROTOR   STATOR   ASSEMBLY

Fig. 14-8   Components and Assembly of Axial-flow Compressor.

There are two types of compressors, axial flow and centrifugal flow. The axial flow compressor is comprised of a series of rotor blades and stator blades. The air moves rearward along the axis of the rotating shaft. The spinning rotor blades scoop the air and force it rearward. The stator blades (stationary) pick up the air and deflect it on to the next rotor blade. The rotor and stator stages get smaller and the angle of the blades scooping the air gets sharper. Air pressures and air acceleration, therefore, increase with each successive compressor stage.

The centrifugal flow compressor consists primarily of three main parts: an impeller, a diffuser, and a compressor manifold. Air is scooped up by the impeller blades as it enters the compressor and is thrown to the outside rim of the impeller by centrifugal force. The impeller is designed in such a way that the speed and pressure of the air increase as the air is thrown outward. As the air comes off the impeller rim, it is picked up by the diffuser and channeled to the compressor manifold. The manifold delivers the air on to the combustion chamber.

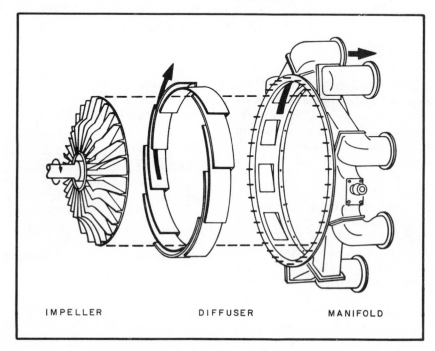

IMPELLER   DIFFUSER   MANIFOLD

Fig. 14-9   Components of a Centrifugal-flow Compressor.

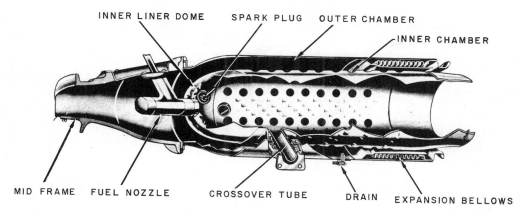

Fig. 14-10   Can-type Turbojet Combustion Chamber.

The combustion section contains the combustion chambers, spark plugs, and fuel nozzles. Within the combustion chambers the air and fuel are burned, the hot gases being used to power the turbine and to produce the required thrust. The size and number of combustion chambers may vary from engine to engine. However, the combustion chambers themselves are usually made up of these parts: outer combustion chamber, inner liner, inner liner dome, flame crossover tube, and fuel injector nozzle.

The outer chamber receives the high-pressure air and surrounds the inner chamber much like a large tin can with a small can inside. The inner liner has many holes in its surface. Air enters through these holes to be burned with the jet fuel. The flame sweeps rearward down the inner liner. Incoming air all along the liner keeps the flame from touching the metal. Combustion is completed just before the hot gases reach the turbine. Ignition is started by spark plugs but, once started, the plugs are turned off and burning is continuous. The flame-crossover tubes carry the flame to the other combustion chambers which are not equipped with spark plugs.

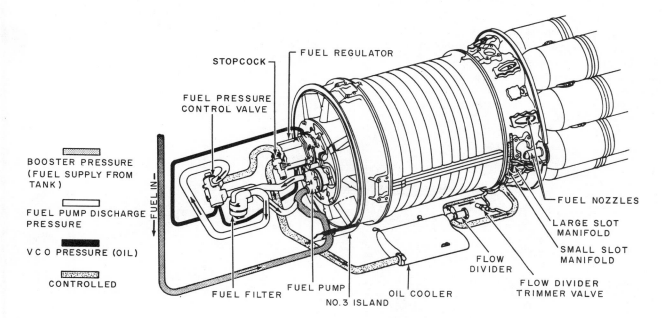

Fig. 14-11   Turbojet Fuel System.

The turbine section harnesses much of the kinetic energy of the hot gases, using the power to drive the compressor and engine accessories. The turbine is basically a dynamically balanced disc with steel alloy blades or buckets fastened to its periphery. A stationary turbine nozzle vane assembly directs the hot gases against the rotating turbine blade. The turbine may be a single rotor or it may be designed with several turbine rotors.

The exhaust section is designed to discharge the exhaust gases in a way that will produce maximum thrust. The turbulent gases

from the turbine are straightened and concentrated by an inner cone and expelled through the tailpipe.

There is always a surplus of oxygen in the jet engine. The oxygen in the air mass does serve to cool the engine parts but otherwise it goes out the exhaust. Some engines are equipped with after-burners to utilize this oxygen and to supply additional engine thrust. The after-burner is behind the combustion and turbine section. More fuel is introduced in the after-burner, increasing the force of the expelled gases.

Fig. 14-12   Turbine Elements.

TURBOPROP

Essentially the turboprop engine consists of a turbojet engine that drives a propeller. There are some differences in the designs of turboprop engines, but the basic engine components are present: inlet, diffuser and duct, compressor, burner, turbine, exhaust duct, and exhaust nozzle. The principles of operation are virtually the same as those of the turbojet and, therefore, will not be repeated.

The engine turbine, however, is designed to extract more power from the hot gases since the turbine must drive the propeller. The propeller is responsible for roughly 90 percent of the engine's thrust, the remaining thrust being produced by the gases leaving the exhaust nozzle. The shaft speed of the jet engine is far greater than the speed range of a propeller; therefore, a reduction gear assembly is used to obtain the desired propeller speed.

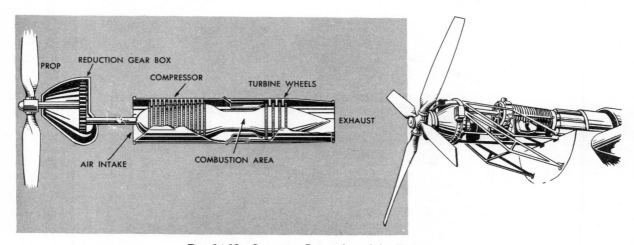

Fig. 14-13   Operating Principles of the Turboprop.

The turboprop aircraft is especially suited for commercial cargo and passenger service where flights are relatively short and turbojet speeds are not essential. Runways may be shorter for these aircraft since the engine's performance is excellent at low speeds. The turboprop combines the power of the turbojet with the propulsive efficiency of the propeller. The turboprop engine can produce about twice as much power as a conventional piston engine of equal weight.

Another variation of the turboprop engine is the turbofan engine. On this type of engine a ducted fan replaces the propeller. The large fan is really the first section of the compressor; however, much of the air is released before the final compressor stages, burner, and turbine. The fan accelerates the air mass much the same as the propeller would.

The fan is driven at shaft speed and is designed to operate more efficiently at higher air speeds than is possible with a propeller. The fan is responsible for 30% to 60% of the propulsive force. The turbofan engine is also lighter and simpler than the turboprop engine.

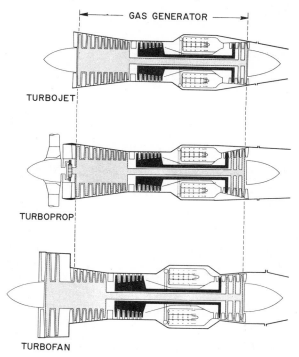

Fig. 14-15   Gas Generator Sections of Turbojet, Turboprop and Turbofan Engines.

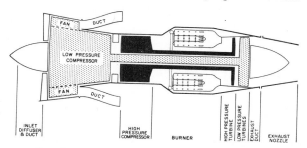

Fig. 14-14   Dual-Compressor Turbofan Without After-burner.

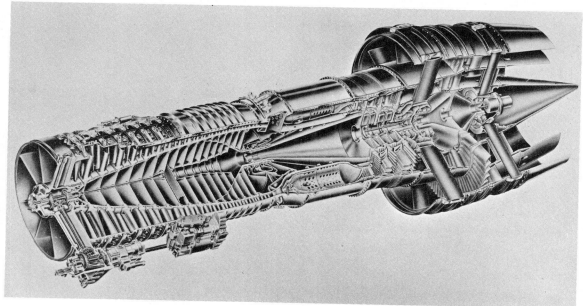

Fig. 14-16   General Electric Turbofan Engine CJ-805-23 with Fan Mounted Aft.

JET FUELS

During the early development of jet engines it was commonly thought that the engine could operate on almost any petroleum fuel. Soon it was found that kerosene was a desirable fuel.

Kerosenes cover quite a range of fuel characteristics so very fine fuel specifications have been laid down for jet fuels.

Most jet fuels are not pure kerosene, but rather a blend of kerosene and gasoline. A common ratio is 65% to 70% gasoline to 30% to 35% kerosene. Military jet fuels are designated by JP-1, JP-2, JP-3, etc. Jet fuel JP-4 and JP-5 are the most common both for military and commercial use. Jet fuel must be high quality and made to meet rigid specifications. Often, jet engines are designed to operate on one specific fuel type.

GENERAL STUDY QUESTIONS

1. Is the theory of jet propulsion quite new or relatively old?
2. What does Newton's third law of motion state?
3. What term is used to describe the propulsive force of a jet engine?
4. How are jet engines and rocket engines alike?
5. How are jet engines and rocket engines different?
6. Does the ramjet engine have any moving parts?
7. Can the ramjet take off on its own power?
8. Does the ramjet have any practical use today? What?
9. Explain the pulsing operation of the pulsejet.
10. Is the pulsejet capable of very high speeds?
11. What two countries took the lead in developing the turbojet?
12. What are the main sections of the turbojet engine?
13. What amount of energy is needed to operate the compressor?
14. What is the purpose of the compressor?
15. What is the purpose of the turbine?
16. Why is the design of the exhaust system important?
17. Does the turbojet burn all the air that it takes in?
18. How is a turboprop engine different from a turbojet engine?
19. What are the advantages of a turboprop engine?
20. What are the advantages of a turbofan engine?

CLASS DISCUSSION TOPICS

● Discuss how a balloon illustrates Newton's third law of motion and jet propulsion.

● Discuss other "everyday" examples of jet propulsion.

● Discuss why the jet engine does not need to push against air to produce thrust.

● Discuss the air-dependent and air-independent engines.

● Discuss the operation of the ramjet, pulsejet, turbojet, turboprop and turbofan.

ROCKET ENGINES

Rockets were first used in thirteenth century China. Hoping to be launched into space, Wan Hu, a scholar and scientist, lashed rockets to his sedan chair. When he lit the rockets, Wan Hu blew up, becoming the first casualty in space flight.

Rockets have been used by military men for several centuries. The Chinese of Wan Hu's time used rockets for warfare, and the Indians used rockets against the British in India in the late eighteenth century. "The rockets' red glare" in our National Anthem refers to the rockets used by the British in the War of 1812. Rockets served as signals and flares in World War I. It wasn't until World War II, however, that rockets were used as missiles on a large scale.

The Germans spent 40 million dollars to develop the V-2 rocket, and about 2,675 of these rockets were used against the Allies in World War II. Though inaccurate, the V-2 rocket could travel up to 200 miles at a speed of 3,600 m.p.h., and it was responsible for much destruction. The V-2 was huge: 47 feet long, 5.5 feet in diameter, and 27,000 pounds at takeoff.

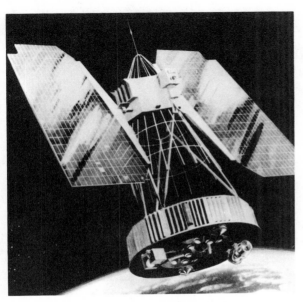

Fig. 15-1 Nimbus I Weather Satellite

In the United States, Dr. Robert H. Goddard was an early rocket pioneer who developed the forerunner of the famous Bazooka of World War II, an antitank missile. His primary interest, however, was the development of high altitude meteorological rockets.

Rocket development gained more and more national attention after World War II. The United States developed the Viking and Aerobee rockets for meteorological research. The Honest John, Redstone, and Nike rockets were designed to strengthen our defense network.

The United States and Russia soon directed their efforts to the intercontinental ballistic missile (ICBM), capable of traveling over 5,000 miles and yet striking a target with pinpoint accuracy. Fortunately for civilization the ICBMs have never been used, but the technology of these rocket weapons has provided a base for peaceful space exploration.

SPACE EXPLORATION

Before space exploration programs could be started, it was necessary to develop an engine with a thrust powerful enough to overcome the earth's gravitational pull. Acceleration is the key to escape, and in the last twenty years rockets have been able to achieve astounding acceleration rates. As they consume their fuel, they become increasingly lighter, and so they can accelerate at higher and higher rates of speed. Staging also aids rapid acceleration. Three stages are assembled for liftoff. When the first-stage propellants are exhausted, the first stage of the rocket drops off. Already at a high velocity and now considerably lighter, the second stage ignites and its acceleration increases. When the second stage burns out, it drops off, and the third stage develops the velocity and acceleration necessary for leaving the earth's gravitational field and traveling into space.

The rocket's path in space is called its trajectory. Four general trajectories are now utilized: sounding, earth orbit, earth escape, and planetary.

Sounding rockets are used for research of the upper atmosphere. They are two-stage rockets which go straight up and then fall back into the ocean or desert. Scientific data is radioed or parachuted back to earth. The Argo, Aerobee, and Astrobe are sounding rockets.

Earth orbit is attained by launching the rocket vertically and then tilting the trajectory so that it becomes parallel to the earth's surface. When the desired altitude (125 miles) is reached, and orbital velocity (18,000 m.p.h.) is attained, the final-stage rocket shuts off. The gravitational pull of the earth and the centrifugal pull of the rocket balance each other, and the spacecraft coasts into orbit around the earth. The earth-orbiting rocket is commonly referred to as a satellite. The first satellite to orbit the earth was the Russian Sputnik 1, launched on October 4, 1957. Explorer 1, launched on January 31, 1958, was the first United States satellite. Explorer 1 weighed 30.8 pounds and orbited the earth every 1 hour 55 minutes. Several hundred research satellites have been launched since that time.

The next goal was to place a manned spacecraft into orbit around the earth and return it safely to the earth. The Russian Vostok and American Mercury programs were directed toward this goal. On April 12, 1961, Russian cosmonaut Yuri Gagarin became the first human to orbit the earth. On May 5, 1961, Alan Shepard Jr. rode a ballistic flight into space and traveled 300 miles downrange on the Freedom 7. John Glenn Jr. was the first American to orbit the earth. He flew the Friendship 7 for three orbits around the earth before returning safely home. Glenn's spacecraft was powered by an Atlas booster rocket.

The United States Gemini Program and the Russian Voskhad Program went beyond the previous one-man flights that proved man's ability to function in space and return to earth. The Gemini Program involved rendezvous with a target vehicle and docking with it. Extreme precision was necessary to accomplish the

rendezvous. The Gemini spacecraft had to be launched into exactly the same orbit as that of the target vehicle which had been launched a day earlier. In addition, Gemini astronauts left their space capsules and walked in space. The Gemini astronauts of 1965 and 1966 accomplished their objectives. Edward White II became the first American to walk in space on June 3, 1965. Astronauts Frank Borman and James Lovell performed a rendezvous between Gemini 6 and 7. The Titan II rocket was the launch vehicle for the Gemini Program. The Titan II was a two-stage rocket,

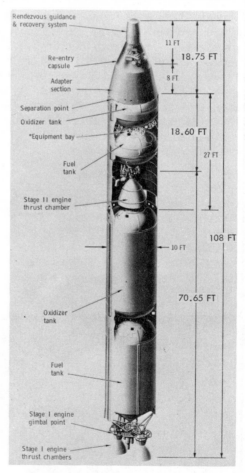

Equipment bay contains:

- Batteries
- Malfunction detection system (MDS) units
- Range of safety command control system
- Programmer
- Three-axis reference system (TARS)
- Radio guidance system (RGS)
- Autopilot.
- Instrumentation and telemetry system

Fig. 15-2 Titan II Launch Vehicle for the Gemini Spacecraft.

109 feet high, delivering 430,000 pounds of thrust at sea level and able to put into orbit the Gemini space capsule, which weighed more than 7,000 pounds.

Earth escape is a step beyond orbital travel. The craft must gain enough velocity (about 25,000 m.p.h.) to escape the gravitational influence of the earth. After its escape, the spacecraft eventually goes into orbit around the sun, like a tiny planet, unless it comes under the gravitational influence of another celestial body. The Apollo Program, using the powerful Saturn 5 launch vehicle, was designed to escape the earth, orbit the moon, land on the moon, and then return to the earth.

Apollo 11 put the first human being on the surface of the moon. The 365-foot, 3-stage

Fig. 15-3 Gemini IV Spacecraft Lifts Off, Powered by a Titan II Rocket.

Saturn rocket began its journey with the ignition of the first stage, burning 214,000 gallons of kerosene and 346,000 gallons of liquid oxygen in 2 1/2 minutes. At the completion of this stage, the rocket was traveling at the rate of 6,100 m.p.h., 38 miles above the earth. The second stage burned liquid hydrogen and liquid oxygen and attained a speed of 15,300 m.p.h. and an altitude of 117 miles. The third stage, using the same kind of fuel, placed the spacecraft in orbit around the earth. While in orbit the spacecraft was checked out thoroughly before the third stage was fired again. A final firing of the third stage accelerated the spacecraft to 24,200 m.p.h., and the trajectory of the craft was changed for the moon.

The spacecraft consisted of three modules: the command module which carried the astronauts, the service module, which consisted of the rocket and fuel tanks, and the lunar module, which was equipped to land and take off from the moon.

As the spacecraft entered the moon's gravitational field, it was slowed down with retro-rockets and put into lunar orbit. Two astronauts transferred from the command module to the lunar module, separated the lunar module from the others, and descended to the moon's surface. Once on the moon, the astronauts collected rocks and performed scientific experiments. Then the lunar module, using its landing section as a launch pad, blasted off at exactly the correct time to rendezvous with the orbiting command and service modules. After docking, the rocket in the service module was burned to power the craft back to earth. The lunar module was left to orbit the moon, and the service module was disconnected outside the earth's atmosphere, leaving only the command module to return to earth. Neil Armstrong, Michael Collins, and Edwin Aldrin accomplished the historic Apollo 11 mission in July 1969. Further Apollo flights, following the same pattern of operations, permitted further exploration and observation of the lunar surface.

The Space Shuttle Program, begun in 1972, is intended to provide a space transportation system in earth orbit, employing a reusable manned spacecraft. It will be able to carry a load of cargo and passengers totaling 65,000 pounds and land on a jet-sized airstrip.

Fig. 15-4 The Apollo/Saturn Vehicle Generates a Thrust of 7½ Million Pounds at Liftoff.

Eventually the space shuttle will carry passengers and cargo between the earth and manned orbiting space laboratories. Like other earth-orbiting spacecraft, it will consist of two stages: the first stage, the booster, lifts the shuttle into an orbital mission; and the second stage, the shuttle, separates from its booster and enters its orbit under its own power.

Fig. 15-5 The H-1 Engine, Generating up to 188,000 Pounds of Thrust, is Used in a Cluster of Eight to Provide the 1,500,000 Pounds of Thrust for the Saturn Booster.

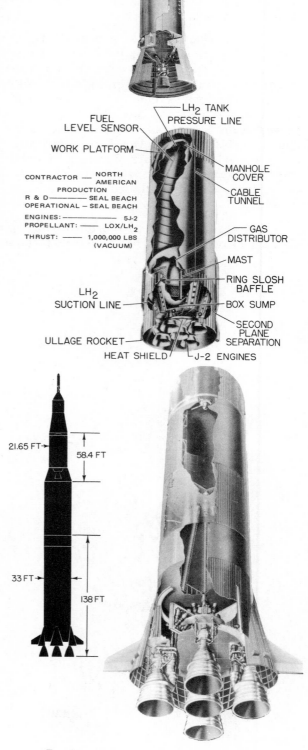

LH₂ TANK
PRESSURE LINE
FUEL
LEVEL SENSOR
WORK PLATFORM
MANHOLE
COVER
CONTRACTOR — NORTH
AMERICAN
PRODUCTION
R & D — SEAL BEACH
OPERATIONAL — SEAL BEACH
ENGINES: — 5J-2
PROPELLANT: — LOX/LH₂
THRUST: — 1,000,000 LBS
(VACUUM)
CABLE
TUNNEL
GAS
DISTRIBUTOR
MAST
RING SLOSH
BAFFLE
LH₂
SUCTION LINE
BOX SUMP
SECOND
PLANE
SEPARATION
ULLAGE ROCKET
HEAT SHIELD
J-2 ENGINES

21.65 FT
58.4 FT
33 FT
138 FT

Fig. 15-6 First, Second, and Third Stages of the Saturn V Launch Vehicle.

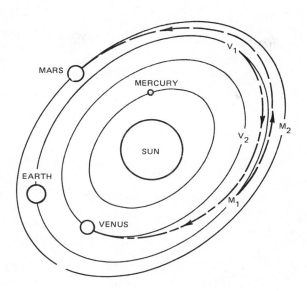

M$_1$ = POSITION OF EARTH WHEN ROCKET IS LAUNCHED FROM EARTH FOR TRAVEL TO MARS
M$_2$ = POSITION OF MARS WHEN ROCKET IS LAUNCHED FROM EARTH FOR TRAVEL TO MARS
V$_1$ = POSITION OF EARTH WHEN ROCKET IS LAUNCHED FROM EARTH FOR TRAVEL TO VENUS
V$_2$ = POSITION OF VENUS WHEN ROCKET IS LAUNCHED FROM EARTH FOR TRAVEL TO VENUS

Fig. 15-7 Transfer Trajectory for Launching
Rockets to Mars and Venus.

Planetary flights are more difficult than lunar flights. The moon is in the earth's gravitational field and is traveling around the sun at the same rate as the earth. The relative orbits of the earth and the planets around the sun make the problem of interplanetary flight complex. Favorable opportunities to launch rockets for Venus and Mars occur about every two years. These periods for favorable launching are called launch windows. Much planning is required to develop a transfer trajectory, which is the elliptical trajectory used for launching a rocket from the earth to another planet, figure 15-7. Both the United States and Russia have launched space probes to other planets. The Mariner Program has probed Venus and Mars. Atlas Centaur and Atlas Agena have powered these scientific flights.

ROCKETS ARE AIR-INDEPENDENT

Rockets are unique as a major power producer primarily because they are air-independent. Rockets operate equally as well whether there is any air present or not. The rocket moves well through the troposphere, stratosphere, ionosphere, or in the vacuum of outer space. Actually, the thinner the air around the rocket, the faster the rocket can travel since there is less resistance from air friction.

The rocket engine burns fuel just as any other engine does, by changing potential chemical energy into useful kinetic energy. However, because it very often must travel in conditions where there is no air present, the rocket can in no way be dependent on oxygen in the atmosphere for maintaining its combustion. Therefore, the rocket carries its own supply of oxygen or oxidizer and, consequently, is air-independent.

Just as in jet propulsion, the thrust of the rocket comes from within the rocket itself. As the fuel ignites, the rapidly expanding gases of combustion rush toward the rear of the rocket. Here again, Newton's third law of motion, "for every action there is an equal and opposite reaction," comes into play. The action of the gases leaving the rear of the rocket creates an equal but opposite reaction of pushing the rocket forward.

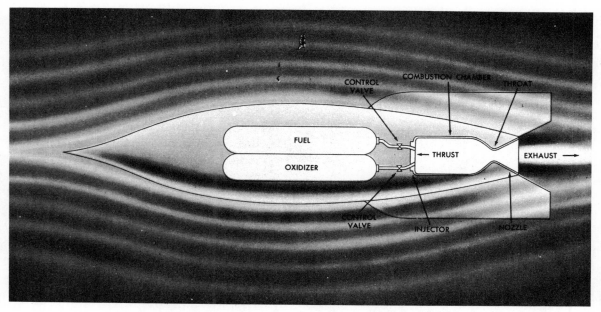

Fig. 15-8 Operating Principles of a Liquid Propellant Rocket

TYPICAL ROCKET OPERATION

It might be helpful to quickly examine the operation of a typical liquid propellant rocket before studying more specific rocket construction.

The propellant, our oxidizer and fuel, is at rest in tanks. The propellant represents untapped chemical energy.

During rocket operation, a feed system forces the propellant from the storage tanks into the rocket engine's injector. This feed system applies great force, greater than the high pressures usually found in the combustion chamber. The injector thoroughly mixes the oxidizer and fuel in the proper proportions.

Ignition begins at the face of the injectors as the propellant enters the thrust chamber. Burning is continuous; the liquids are converted into hot burning gases that travel back to the nozzle and out of the rocket.

Most of the combustion takes place before the gases reach the nozzle but some combustion continues even as the gases leave the rocket.

Most rocket engines use the convergent-divergent (C-D) nozzle for accelerating the gases and producing the proper thrust. As the name suggests, the cross-sectional area of the nozzle is at first small, diverging to a larger cross-sectional area at the rear of the engine.

The speed of the burning gases in the combustion chamber accelerates to 4000 feet per second as they travel rearward in the convergent section. At the throat of the nozzle, the gases will attain the speed of sound (Mach 1). This "local speed of sound" is not 760 m.p.h. but the speed corresponding to the temperature at a certain point in the C-D nozzle.

In the divergent section of the C-D nozzle the gases continue to expand, increasing the velocity to 6,000 to 8,000 feet per second, supersonic speed. The diverging section of the nozzle actually increases the thrust produced. A simile of the divergent section's action would be a balloon which is forced into the large end of a funnel. When released, the balloon moves away exerting pressure to "thrust" the funnel forward.

Rocket engines are classified into two major groups:

(1) Liquid Propellant

(2) Solid Propellant

Generally, liquid propellants are used where greater impulse and range are necessary, such as Intercontinental Ballistic Missiles.   Solid propellants are used for short range missiles.

LIQUID PROPELLANT ROCKETS

The liquid propellant rocket may have a Monopropellant which is a mixture of two or more compounds. The oxidizer and fuel are blended together to form a single substance. They, therefore, offer the advantage of requiring only one storage tank and one feed system. An example of a monopropellant would be hydrogen peroxide and alcohol mixed together.

A rocket may have the Bipropellant system which has the oxidizer and fuel stored in separate tanks. This propellant system requires two storage tanks and a more complicated feed system. However, the bipropellant system has been more successful than the monopropellant system and currently is in wide use. A common system would use liquid oxygen and alcohol.

The feed system's job is to force the liquid propellant into the combustion chamber. The pressure feed system is relatively simple; it uses gas under high pressure to displace the oxidizer and fuel.

A tank of high-pressure gas is led through the necessary accompanying pressure regulator, feed lines, and control valves and into the propellant tanks. Propellant is delivered to the combustion chamber at the proper rate.

This system is simple, trouble free, and reliable but there is one notable limitation: the weight of the high-pressure gas tank is too great for many applications. However, the pressure feed system is seen on assisted take-off units and on rocket sled engines.

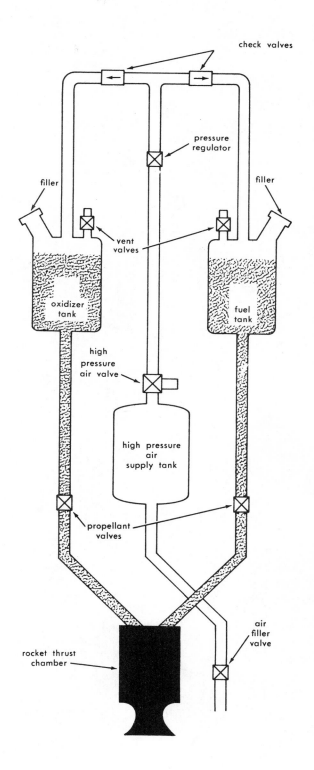

Fig. 15-9 Gas Pressure Feed System for Liquid Propellant Rockets

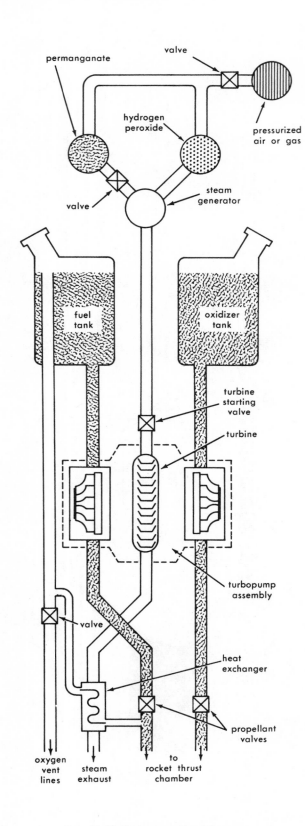

Fig. 15-10 Turbopump Feed System

The pump feed system uses a centrifugal pump to deliver the propellant. The pump is driven by a turbine and the system is commonly referred to as the turbopump feed system. The turbine may be driven by several systems:

(1) The gas bleeding or "bootstrap" system takes a portion of the hot, high-pressure gases from the main combustion chamber, cools them somewhat, and then uses this power to drive the turbine.

(2) The gas generator generates high-pressure gases by a chemical reaction of the propellants. The reaction is similar to that in the main combustion chamber but temperatures are lower. The V-2 rocket used this type of generator to drive its turbine; potassium permanganate and hydrogen peroxide were sprayed together, the high temperature gases produced by the chemical reaction being used to drive the turbine.

(3) The high-pressure gas system is equipped with storage tanks filled with high-pressure air or nitrogen. The controlled flow of the gas from these tanks is directed against the turbine blades which drive the turbopump.

A gas generator drives the centrifugal pump which delivers the propellant to the injectors and thrust chamber.

The large, liquid propellant rocket engine for use in the Atlas, Thor, Jupiter, and Redstone missile programs consists primarily of a propellant feed system, control system, and thrust chamber.

In Fig. 15-10 can be seen the small spherical gas generator, the power source for the high performance turbopump. The latter rotates twin pumps forcing fuel and oxidizer through the injector into the combustion chamber. The thermal energy of expanded gases resulting from the combustion of propellants within the chamber is converted to direct kinetic energy as the gases are expelled through the nozzle to produce thrust.

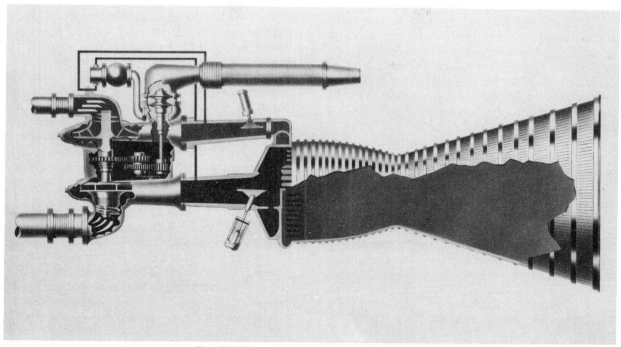

Fig. 15-11 Liquid Propellant Rocket Engine

SOLID PROPELLANT ROCKETS

The principal action of a solid propellant rocket is the same as that of a liquid propellant rocket. Both produce thrust as hot gases are discharged through the rocket engine's nozzle. Solid propellant rockets are inherently simpler in design and construction, present fewer servicing problems, and are believed to be more reliable than liquid propellant rockets. Since they generally produce thrust for a relatively short time, they are used extensively for missile and aircraft boosters, guided missiles, and rocket projectiles.

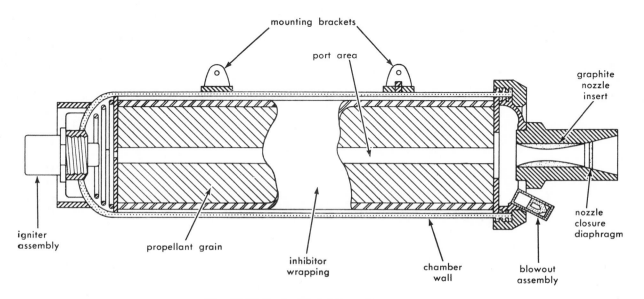

Fig. 15-12 Typical Solid Propellant Rocket

The propellant (fuel and oxidizer) is usually made up of a type of hydrocarbon and an oxidizer that has a high weight-percentage of oxygen. The two are mixed, forming a liquid or plastic mass which is poured into a mold and allowed to solidify or set. In some cases the rocket casing is the mold. The finished unit of propellant is referred to as a "grain".

One of the first solid propellants used was black powder. Carbon and sulphur (fuel) and saltpeter (oxidizer) were mixed with glue or oil (binder) to form a mass that could be shaped, and then allowed to solidify. A common propellant is Ballistite; it has two chemical bases, nitroglycerine and nitrocellulose. Galcit, composed of an asphalt-oil mixture and potassium perchlorate, is another solid propellant.

The rate of burning, thrust, operating pressure, and duration of burning are determined by the construction of the grain, its size, shape, and exposed burning surface. The burning rate of a propellant is expressed in "inches per second", the number of inches burned through the propellant each second measured perpendicular to the burning surface. Most common propellants burn at a rate between 0.03 and 2.5 inches per second.

Solid propellant rockets are broadly classified as restricted burning and unrestricted burning. Generally, the restricted burning propellant will deliver thrust for a longer time while the unrestricted burning propellant will deliver more thrust but for a shorter time. The restricted burning propellant has some of the exposed propellant surface covered with an inhibitor which controls the burning rate. The unrestricted propellant allows all of the exposed surfaces to burn at the same time. Thrust and time range of typical restricted burning grains of propellant are 100 to 10,000 pounds thrust for 4 to 120 seconds. The range for typical unrestricted burning grains of propellant is 500 to 100,000 (plus) pounds thrust for 0.05 to 10 seconds.

Solid propellants may be molded into different shapes in order to produce the desired burning characteristics. Progressive burning means that more surface is exposed as the propellant burns, thus increasing thrust. Regressive burning means that less surface is exposed as the propellant burns, decreasing thrust. Neutral burning means that the same amount of surface is exposed as the propellant burns maintaining constant thrust.

Rocket engine power may never become a power unit for the average consumer to use but it has already written a sizable chapter in our recent history. Our nation's defense rests more and more on weapons powered by rocket engines. The present and future exploration of outer space and other planets is and will be carried on with the aid of rockets that were literally "unthinkable" just a few years ago.

GENERAL STUDY QUESTIONS

1. Briefly discuss the rocket's early history.

2. What country was responsible for the V-2 rocket, the forerunner of modern rockets?

3. List some modern uses for rockets.

4. Explain the term "air-independent".

5. What are the two major classifications of rockets?

6. Explain the terms, monopropellant and bipropellant.

7. What is the function of the rocket feed system?

8. What are the advantages of a solid propellant rocket?

9. What is the solid propellant rocket's major limitation?

10. What is the difference between restricted and unrestricted burning solid propellants?

CLASS DISCUSSION TOPICS

- Discuss how the work of Dr. Goddard and German rocket experts has laid the foundation for modern rockets.

- Discuss the peaceful and military weapon use of rocket engines.

- Discuss the air-independent operation of a rocket.

- Discuss the flow of propellants and combustion gases through a liquid propellant rocket.

- Discuss the advantages and disadvantages of both solid and liquid propellant rockets.

Unit 16

GAS TURBINE ENGINES

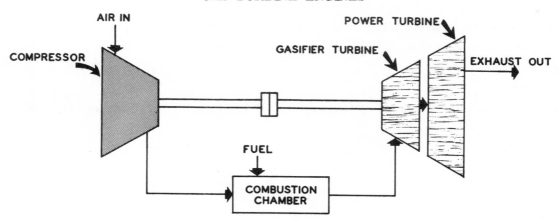

Fig. 16-1 Simple Open Cycle Gas Turbine.

The basic principle undergirding the gas turbine engine is that of converting the kinetic energy of hot expanding gases into mechanical rotary energy of a turbine wheel. Essentially, the typical gas turbine first brings large quantities of air into the engine with the aid of a compressor. Then, fuel enters the combustion chambers where it is combined with the air and ignited. The hot, expanding gas media is directed against the buckets or blades mounted on the turbine wheel. In addition to the gasifier turbine that drives the compressor, there is a power turbine that provides shaft power to the driven machine. It is at the turbine wheel that a portion of the kinetic energy of the expanding gases is converted to mechanical rotary energy. An operating cycle of this type is referred to as a simple open cycle because the gases are exhausted directly to the atmosphere, Fig. 16-1.

It is in the area of exhausting gases that much heat and efficiency can be lost in any engine. The waste involved in the reciprocating piston engine is obvious; hot gases are removed from the engine even though much heat energy remains. Gas turbines are frequently designed to minimize this type of waste. Passing the gases through several stages of turbine wheels utilizes the energy to a greater degree than a single turbine wheel. Also, a regenerator that uses exhaust heat to preheat the incoming air increases efficiency, Fig. 16-2.

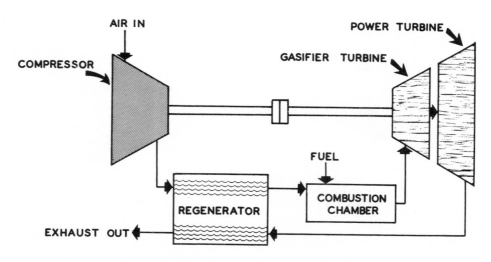

Fig. 16-2 Regenerative Open Cycle Gas Turbine.

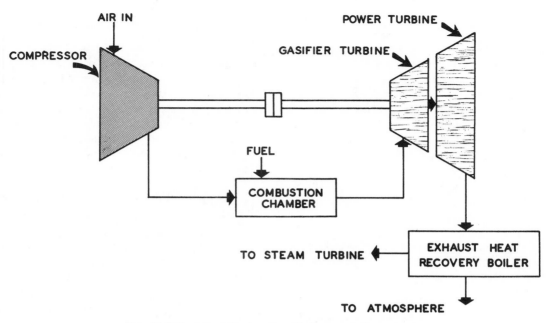

Fig. 16-3 Combined Cycle - Gas Turbine with Steam Turbine.

Another technique that can be used to increase thermal efficiency is to exhaust the gases from the gas turbine to a steam turbine system forming a combined cycle, Fig. 16-3.

Gas turbines can also operate on the closed cycle principle, Fig. 16-4. In this case, hot combustion gases superheat, but do not contact, the working media which is expanded through the gasifier and power turbines, then channeled to the regenerator where it gives up heat, and then to the cooler for further temperature reduction. The media is then compressed and ready for expansion through the system. Since the working media (air or one of several gases) does not come in direct contact with the combustion products, friction losses are reduced as well as turbine wear. The incoming combustion air also has benefit of the regenerator. Any fuel may be used: oil, gas, or solid.

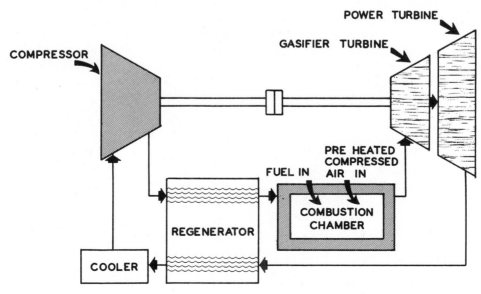

Fig. 16-4 Closed Cycle Gas Turbine.

WHY "GAS" TURBINE?

The term "gas" refers to the media that is produced to drive the turbine wheel. Various fuels can be used such as natural gas, naphtha, distillate oil, residual fuels, jet fuels, diesel fuel, and kerosene. Gas, then, is the combustion product of fuel and air that rushes through the turbine — potential chemical energy converted into kinetic thermal energy.

The thermodynamic cycle of the gas turbine bears similarity to other types of fuel-burning engines. Air and fuel are brought together (ignited) and the resulting hot gases are directed against a surface that can respond to the pressure, surfaces such as a turbine blade or a reciprocating engine's piston. Exhaust gases are exited from the engine. However, the turbine is a "steady flow" process while reciprocating engines are a "batch" process; a batch of fuel mixture enters on the intake stroke and is processed as a unit.

Gas turbines have a hefty appetite for air which can be satisfied only by drawing air into the turbine with a compressor. A portion of the kinetic thermal energy must be used by the gasifier turbine to drive the compressor. The compressors themselves may be axial flow, centrifugal flow, or a combination of both.

GAS TURBINES AND JET ENGINES

This introduction may bring to mind the jet engine principles and components that were discussed in Unit 14, Jet Engines, and it will be well to clarify the distinction that has been drawn between the two prime movers. To be sure, the turbojet discussed in the earlier unit is clearly a turbine engine, but the primary propulsive power of the engine as applied to the aircraft is that of the propulsive reaction of the gases producing thrust as they leave the engine. And, even though the vast portion of the power produced by a turboprop aircraft is used to power the propeller, still about 10% of the propulsive force is derived from the jet principle as gases leave the engine.

The gas turbines discussed in this unit do not rely on jet thrust principles. Their construction and application is strictly that of a turbine shaft driving a mechanism.

GAS TURBINE APPLICATIONS

Gas turbines have found many uses although their intense development, like that of the turbojet, has been underway for a relatively short time. In some applications their use has been rather firmly established such as in electrical generation and in powering helicopters. However, in other applications, such as passenger automobiles, research and development continue.

Applications of turbines can be classified into several fields:

1. Gas turbines for trucks, buses, large military vehicles, and off-the-road equipment.

2. Gas turbines for automobiles.

3. Gas turbines for helicopters and small propeller aircraft.

4. Gas turbines for marine use.

5. Gas turbines for process industries.

6. Gas turbines for electrical generation.

7. Gas turbines for multipurpose use.

ADVANTAGES OF GAS TURBINES

Gas turbines can cite many strong points. High on the list is the smooth, vibration-free power that is produced. There are no reciprocating parts — just continuous rotary motion. Characteristically, gas turbines require little warmup time; they can be brought up to full operating power in a short time. Of course, advantages vary with the application but it could be further said that gas turbines

1. Compete very well with other prime movers in size and weight per hp.

2. Operate well on a wide variety of relatively inexpensive fuels.

3. Present no significant maintenance problems and, in fact, are considered to be easy to maintain.

In the case of vehicular applications, there are additional advantages. Gas turbines

1. Are air-cooled; no radiator freezing problems.

2. Have fewer moving parts than reciprocating engines.

3. Start very well in cold weather.

4. Combustion is very complete; exhaust is cleaner than that of other engines and therefore the gas turbine represents less of an air pollution problem.

5. Will not "stall" under overload.

6. Heat for passengers is instantly available.

SPECIAL CONSIDERATIONS

Gas turbines do require some special considerations in certain areas. One is their appetite for air: a compressor is needed and the air must be well cleaned or it will damage the precision parts within the turbine. And, turbines operate best at top speed. To slow a turbine down by reducing the fuel available lessens the efficiency of the turbine somewhat. These problems have been a real hurdle for engineers and researchers to jump, especially in vehicular applications.

ENGINE SIZE COMPARISON
GAS TURBINE vs. CONVENTIONAL

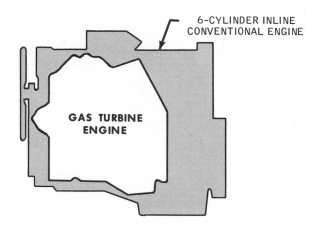

VEHICULAR DEVELOPMENT

Vehicular turbine applications must answer the problems of fuel consumption, wide variety of speed requirements, exhaust temperature, and "engine braking." The regenerative gas turbine is employed to answer these problems effectively.

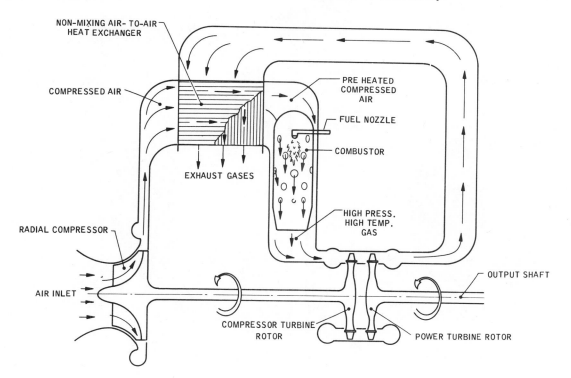

Fig. 16-5 Fluid Flow Through a Regenerative Gas Turbine.

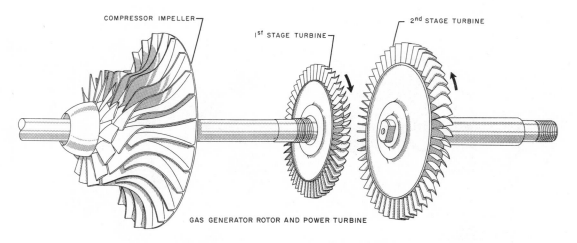

Fig. 16-6 Basic Parts of the Two-Shaft Turbine.

The turbine developed by Chrysler engineers is a two-shaft regenerative type. Fig. 16-6 illustrates the first stage gas generator rotor (gasifier turbine) which drives the compressor and the accessories which are a part of the engine itself and the second stage turbine or power turbine which provides power for propelling the auto. The regenerators are rotating heat exchangers that extract heat from the exhaust gases and then use this heat to preheat air as it comes from the compressor to the burner or combustion chamber, Fig. 16-7. The regenerators serve to increase fuel economy and reduce temperatures.

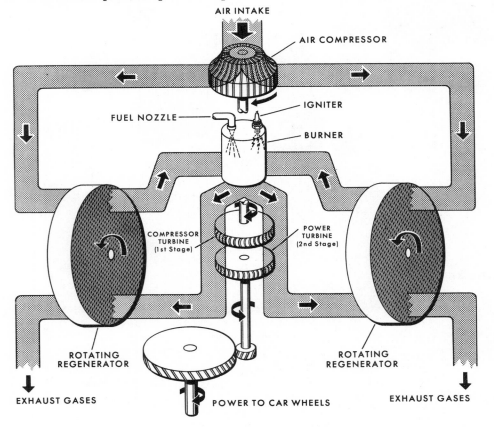

Fig. 16-7 Diagram of Gas Turbine Operation.

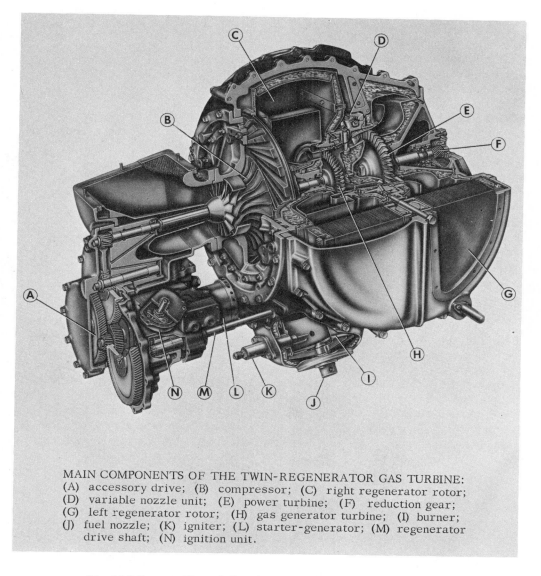

MAIN COMPONENTS OF THE TWIN-REGENERATOR GAS TURBINE:
(A) accessory drive; (B) compressor; (C) right regenerator rotor;
(D) variable nozzle unit; (E) power turbine; (F) reduction gear;
(G) left regenerator rotor; (H) gas generator turbine; (I) burner;
(J) fuel nozzle; (K) igniter; (L) starter-generator; (M) regenerator
drive shaft; (N) ignition unit.

Fig. 16-8 Cutaway View of Chrysler Regenerative Gas Turbine A-831.

Air is drawn into the turbine and compressed by a centrifugal compressor. The compressed air picks up heat as it passes through the regenerator. Arriving at the burner, the air is burned with the fuel. Through the combustion process, the energy level of the fuel is increased greatly and this media is applied to the first stage turbine which drives the compressor. The hot gases are then applied to the second stage turbine or power turbine which drives the auto through an appropriate transmission. Before entering the exhaust ducts, the hot gases pass through the honeycomb regenerator drums.

Naturally, the first stage turbine always rotates whenever the turbine is operating, at speed of 18,000-22,000 r.p.m. at idle, moving up to 44,600 r.p.m. at rated power of 130 hp. The power turbine speed may range from 0 r.p.m. when the auto is standing still, to a maximum of about 45,700 r.p.m.

Fixed nozzles direct the hot gases against the first stage turbine wheel. However, at the second stage or power turbine wheel, the variable nozzle is used, Fig. 16-9. This is a feature of the Chrysler turbine. The variable nozzles permit engine braking and high performance at all speed ranges. The nozzles are linked to the accelerator, permitting the proper nozzle direction. At starting or idle, the nozzles are in an axial position, but as the accelerator pedal is depressed, the variable nozzles are turned to a point more and more in the direction of turbine wheel rotation. The variable nozzles are also used to provide engine braking. If the auto is traveling at more than 15 m.p.h. and the accelerator is released, the nozzles are moved to such a position that the gases will be directed against the turbine wheel's direction of rotation. At speeds of less than 15 m.p.h., release of the accelerator will cause the nozzles to assume an axial position or open idling position.

Fuel control must be variable to insure a high degree of operating economy. Operating at full power for all auto speed requirements would be quite wasteful. This fuel control system consists of a fuel pump, governor, pressure regulator, and metering orifice. At constant speeds, the governor will provide fuel to the burner in response to the position of the accelerator. Fuel flow will shut off when the accel-

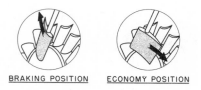
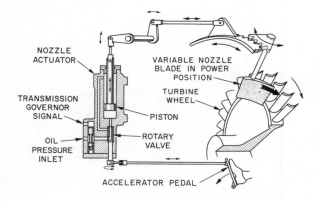

Fig. 16-9 Chrysler Variable Nozzle System.

erator pedal is released, but fuel will resume flowing when idle speed is reached.

This engine has excellent acceleration and flexibility as shown by its horsepower and torque curves. It is tremendously difficult to stall this turbine because even if torque requirements become very great, the turbine will merely slow down; the compressor will continue to function normally.

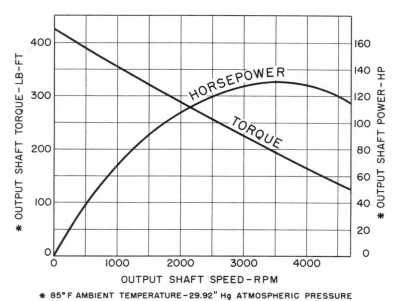

Fig. 16-10 Horsepower and Torque Curves for Chrysler Turbine.

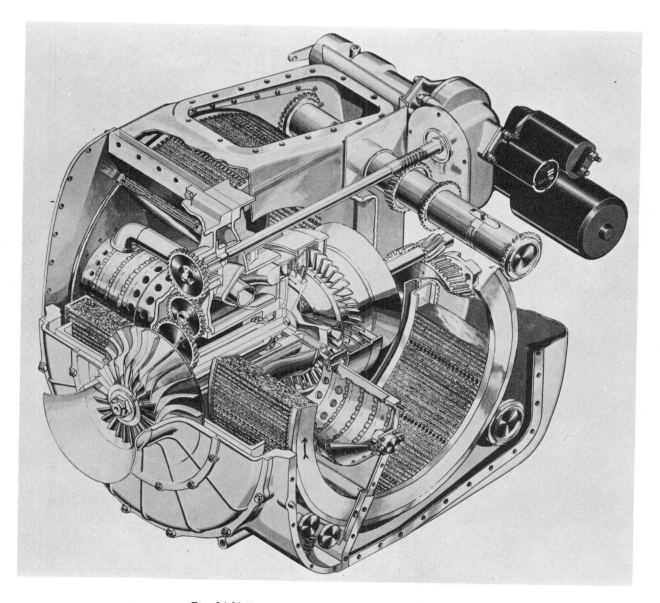

Fig. 16-11 Cutaway of General Motors GT-305 Engine.

General Motors in their GT-305 engine also uses the regenerative principle. The term "regenerative" refers to the engine's design in that the exhaust gases are used to heat the incoming air. This application of heat to the incoming air greatly increases the efficiency of the engine. The principal parts of the engine are: (1) Compressor, (2) Regenerative Drums, (3) Combustors (Burners), (4) Gasifier Turbine, and (5) Power Turbine.

Air enters the radial flow compressor and is compressed to over three atmospheres.

The compressed air goes through the regenerative drums, from the outside surfaces to the inside surfaces, picking up heat as it moves. The preheated air goes into the combustor where it burns with the fuel which is sprayed in. Hot expanding gases are directed against the gasifier turbine by the turbine nozzles. The gasifier turbine is used to drive the compressor. The hot gases next strike the power turbine blades. The power turbine drives the vehicle. Finally, exhaust gases from the turbines pass through the regenerative drums, giving up most of their heat before leaving the engine.

Fig. 16-12 Radial Flow Compressor Compresses the Incoming Air.

The radial flow compressor, Fig. 16-12, rotates at 33,000 r.p.m. Its high-pressure air travels through the regenerative drums, Fig. 16-13, and on to the combustor, Fig. 16-14. A wall within the engine divides the engine into two pressure areas, high-pressure plenum and exhaust plenum. The combustor and about one-third of the regenerative drums are located in the high-pressure plenum. Each drum is a porous matrix about two inches thick. The regenerative drums rotate at 30 r.p.m., picking up exhaust heat in the exhaust plenum and then releasing it as the drum rotates through the high-pressure plenum. It might be said that the drums rotate through the "wall." The pressure areas, drum, combustor, and wall or bulkhead can be seen in Fig. 16-15.

Fig. 16-13 Porous Regenerative Drums Preheat the Air Supply.

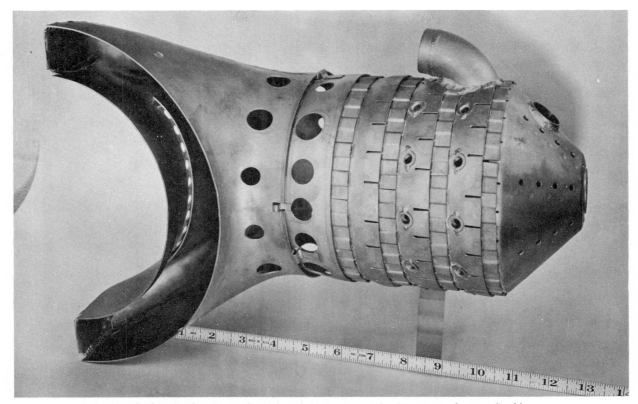

Fig. 16-14 The Combustor (Right) and the Turbine Inlet Transition Section (Left).

Fig. 16-15 Installation of the Combustor.

After the incoming air is passed through the heated drum, it attains a temperature of about 1200°. This superheated air is drawn into the combustor where it is burned with the fuel. Temperatures of about 1650 degrees are reached in the combustor. The hot gases expand through the turbine nozzle assembly and strike the gasifier turbine which rotates the compressor. The gasifier turbine and the power turbine are located between the engine's two pressure areas.

The power turbine which drives the vehicle is not connected with the gasifier turbine. The power turbine rotates at 24,000 r.p.m. (full power), delivering its power to the engine's output shaft. A single stage helical reduction gear reduces the 24,000 r.p.m. to 3,500 r.p.m. Hot gases from the turbine enter the exhaust plenum to give up their heat as they pass through the regenerative drums and out of the engine. Exhaust is discharged at a temperature of 275 degrees at idle or 520 degrees at full power.

Currently, much of the turbine emphasis at General Motors has shifted to applications in the heavy-duty vehicle market, applications that compete directly with diesel engines which now dominate this field. The GT-309 turbine engine has been developed by General Motors to satisfy this need.

Fig. 16-16 General Motors GT-309 Turbine Engine for Heavy Duty Vehicular Applications.

General operating principles for this regenerative gas turbine are similar to those of like power and application. A centrifugal compressor compresses air to 3.8 times atmospheric pressure and directs it through the regenerative drum and to the combustors. High energy gases from the combustion process are directed against the gasifier turbine and then against the power turbine.

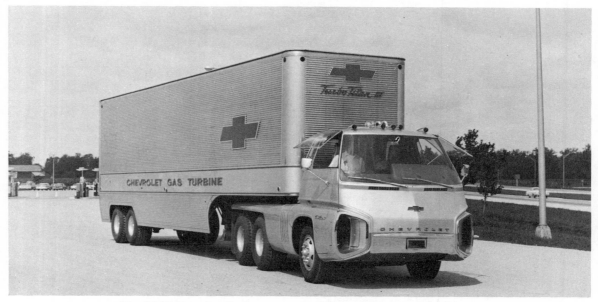

Fig. 16-17 Note the Air Scoops on this GM Truck Turbine Installation.

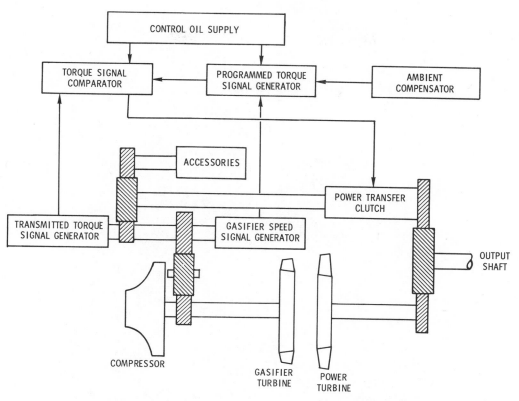

Fig. 16-18 GT-309 Power Transfer Engine Control System Diagram.

The General Motors turbine employs a power transfer system, Fig. 16-18, to insure high efficiency at part throttle and to provide for engine braking. The independent gasifier turbine and the power turbine shafts are equipped with a multiplate slipping clutch which is the heart of the power transfer system. At part load, a portion of the gasifier turbine power is transferred to the power turbine side of the engine. At less than full load, the system would tend to slow the power turbine as a slight amount of pressure is applied to the clutch. The governor on the gasifier turbine will respond by increasing fuel output to increase speed. The net result will be increased turbine inlet temperature and the higher turbine efficiency that can be gained with high temperatures. With power transfer, the gasifier turbine temperature can be maintained at a constant high level, temperatures that will permit high thermal efficiency. Clutch pressure is controlled by a pressure programmer which senses gasifier turbine torque and speed as well as temperature and altitude.

Engine braking is accomplished by locking the power transfer clutch which causes the output of the power turbine to drive the compressor, dissipating the heat energy as air is pumped through the engine.

Fig. 16-19 Heavy Duty Application of the Ford Gas Turbine.

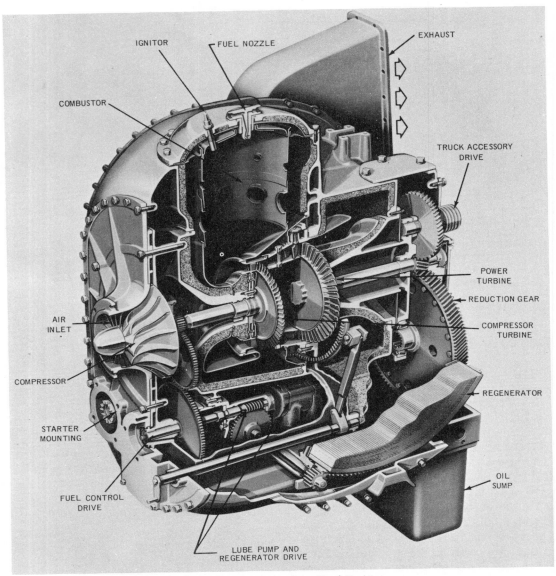

Fig. 16-20 Cutaway of the Ford Turbine.

The gas turbine developed by the Ford Motor Company is in production for marine and industrial applications. Plans are underway for using the turbine in heavy-duty trucking. Ford anticipates that its turbine will compete very favorably in economy and will offer distinct advantages by reducing a truck's weight by approximately 2000 pounds, hence permitting a gain in payload. The turbine will be able to pull an 80,000-pound load at 70 m.p.h. on the modern interstate highway system. Current plans call for the engine and its heavy-duty truck, the W-1000 series, to be in production by the mid 1970s.

The Ford turbine is a low-pressure, regenerative gas turbine engine, developing a maximum of 525 horsepower in its largest series, Fig. 16-20. Its basic operating principles are very similar to other regenerative gas turbines. Maximum compressor speed is 37,500 r.p.m. and the maximum power turbine speed is 31,650 r.p.m. The output shaft of the turbine has a maximum speed of 3000 r.p.m. The gases impinging on the power turbine wheel are modulated in velocity and angle to satisfy various speed requirements. A method similar in principle to that used by Chrysler is used to vary the nozzle angle.

AIRCRAFT DEVELOPMENT

Gas turbines revolutionized the field of commercial and military aviation with turbojet and turboprop engines. And it is not unnatural that gas turbine manufacturers should look to developing gas turbine engines for smaller fixed-wing, propeller-driven aircraft and helicopters — applications where shaft power is utilized rather than jet thrust. The Allison Model 250 turboshaft engine falls into this category. The Model 250 was designed specifically for the new Army light observation helicopter. The 250-C18 weighs a mere 136 pounds, yet can develop 317 hp. and the engine dimensions are a compact 40″ long × 22 1/2″ high × 19″ wide. The power output shaft is geared down to 6000 r.p.m.

The incoming air is compressed by a six-stage axial flow and one-stage centrifugal flow compressor. Air is ducted to the other end of the engine and into the single large combustor. Gases pass through a two-stage gasifier turbine which drives the compressor, then to a two-stage power turbine which delivers power to the output shaft. Exhaust gases are ducted from the engine upward through twin ducts.

In the future, engines such as these may be equipped with regenerative devices which will, of course, increase the engine's thermal efficiency and the potential applications for the engine.

Fig. 16-21 Allison Model 250 Turboshaft Engine.

Fig. 16-22 Cutaway of the AVCO Lycoming T-53 Gas Turbine.

One of the outstanding universal purpose turbines that has been developed is the T-53 by AVCO Lycoming. Depending on the gearing, accessories, and hardware used, the basic T-53 turbine can be used for turbofan, turboprop, geared helicopter, high-speed helicopter, marine, surface vehicle, or industrial applications.

The T-53-L-13 is a shaft turbine helicopter engine that develops 1400 shaft horsepower at 6300 r.p.m., yet weighs only 530 pounds. The overall dimensions of the engine are 47.61 inches long with a diameter of 23 inches. The AVCO AGT-1500 is a vehicular version that develops 1500 brake horsepower at 3000 r.p.m. Again, size is quite small, height is 28 inches, length is 60 inches, width is 40 inches, and weight is 1600 pounds.

The turbine has a five-stage axial compressor followed by a single centrifugal compressor, Fig. 16-22. The compressed air is directed to the rear of the turbine, "around the corner" and to the combustion chamber. Fuel atomizers ring the combustion chamber and, after initial spark ignition, the burning is con-

tinuous. Gases pass through two compressor turbines and then to two power turbines. The power turbine shaft extends forward and concentrically through the compressor shaft, thus delivering shaft power at the forward or cold end of the engine.

ELECTRICAL POWER GENERATION

The field of large horsepower prime movers has long been dominated by the steam turbine; however, the gas turbine has in recent years been seeking to carve its place in this important field. Large, heavy-duty gas turbines may range in size from several thousand horsepower to more than 20,000 horsepower. The prime applications of these turbines is electrical power generation where, depending on the installation of the turbine, they may be used in base load generation or in "peaking." Peaking refers to supplying the needed generation requirements during times of heavy demand on the system. In this application, the base load may be delivered by a conventional steam generator with a smaller gas turbine supplying peaking requirements in a very efficient and economical manner. The gas turbine brings its advantages of quick start and low maintenance into play. Base load installations may be of a combined nature, steam turbine and gas turbine, using the gas turbine exhaust gases to preheat the air before it enters the boiler. Such a combined cycle may increase thermal efficiency as much as 9%.

OTHER APPLICATIONS

Large gas turbines can be used as the prime mover to supply the many phases of an industrial complex, providing electricity, hot water, and heat for the various industrial processes of the industry. By integrating the gas turbine into the process industries such as chemicals, petroleum, petro-chemical, paper, steel, and others, a complete power package can be provided. These industries require large amounts of heat and power and are natural targets for gas turbine installations.

Mobile gas turbine power plants can be prepackaged at the factory and quickly set up at any site that has available fuel.

GENERAL STUDY QUESTIONS

1. Explain how the gas turbine's appetite for air is satisfied.

2. List several fuels that gas turbines are capable of utilizing.

3. Explain the difference between a turboshaft engine and a turbojet engine.

4. Can a gas turbine use both turboshaft and turbojet principles? Explain.

5. List several applications for gas turbine engines.

6. Explain several of the advantages of gas turbine engines.

7. Explain the regenerative principle and the advantage of its use.

8. What is the function of the gasifier turbine?

9. What is the function of the power turbine?

10. Describe the systems that can be used to insure efficiency of the gas turbine at slow speeds and also provide for engine braking.

11. Explain how a gas turbine can be used in a combined cycle with a steam turbine.

CLASS DISCUSSION TOPICS

● Discuss the similarity between the thermodynamic cycle of a gas turbine and other heat engines.

● Discuss the problems and the many steps involved in developing a "new" engine, remembering the years of evolution that the conventional reciprocating gasoline engine has gone through.

● Discuss the importance of utilizing as much of the heat of combustion as possible in the gas turbine engine or in any other heat engine.

● Discuss the advantages of a gas turbine engine.

● Discuss the flow of air, fuel, and hot gases through the engine.

ROTATING COMBUSTION ENGINES

A new prime mover is clearly becoming a member of the internal combustion engine family. Its name: the <u>rotating combustion engine</u>. The key word is <u>rotating</u> versus <u>reciprocating</u>. In a rotating engine the applied heat energy moves the "piston" around in one direction only, not back and forth as in a reciprocating engine.

For years, engineers have recognized the advantages of smooth rotary motion over reciprocating motion. Obviously, in a reciprocating internal combustion engine, a great deal of power is lost in overcoming the inertia of the piston at both the top and the bottom of its stroke when the motion must be stopped and the direction reversed. This power loss problem is one reason why modern engines use lightweight pistons and connecting rods. And smooth as a properly-tuned reciprocating engine may seem, there is a great deal of vibration inherent in its design.

If these facts are true, why then are reciprocating engines "king" and rotating combustion engines, the newcomer? From the times of early scientists and engineers, both rotating and reciprocating principles have been investigated. The reciprocating principle presented fewer manufacturing problems and its geometry or design was easily understood. It caught on and for some eighty years, work toward its perfection has been intensely continued throughout the industrialized nations of the world. The reciprocating engine became the giant and found many uses: reciprocating steam engines, reciprocating diesel engines, reciprocating gasoline engines.

The development of steam turbines (and diesels) have all but eliminated the reciprocating steam engine from the list of modern prime movers. In the commercial aircraft industry and in military aviation, the gas turbine or jet engine has taken over the role of the piston aircraft engine. Gas turbines are also making some inroads in applications for stationary and vehicular power plants. Rotating shaft turbines have made notable progress and are firmly established.

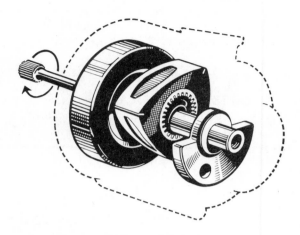

Fig. 17-1 The NSU-Wankel Combustion Engine Utilizes Rotary, rather than Reciprocating, Motion.

However, rotating combustion engines are not turbines that employ hot gases directed against the vanes of a spinning wheel. Rotating combustion engines operate on the basic cycle of <u>intake</u>, <u>compression</u>, <u>power</u>, and <u>exhaust</u> occurring in a pattern that repeats itself.

DEVELOPMENT OF ROTATING COMBUSTION ENGINES

Many persons have experimented with and tested variations of rotating combustion engines down through the years.

1799 Murdock used a Pappeneim gear pump as a single rotating engine.

1846 Galloway invented and built a rotary combustion engine.

1860 Oldam/Franchot invented an internal axis rotary piston compressor.

1882 Parsons designed and built a secondary rotating piston steam engine.

1900 Cooley invented and produced an internal axis single rotation engine with epitrochoidal rotor.

1943 Millary patented an internal axis single planetary rotation machine in which the inner rotor had hypotrochoidal contours.

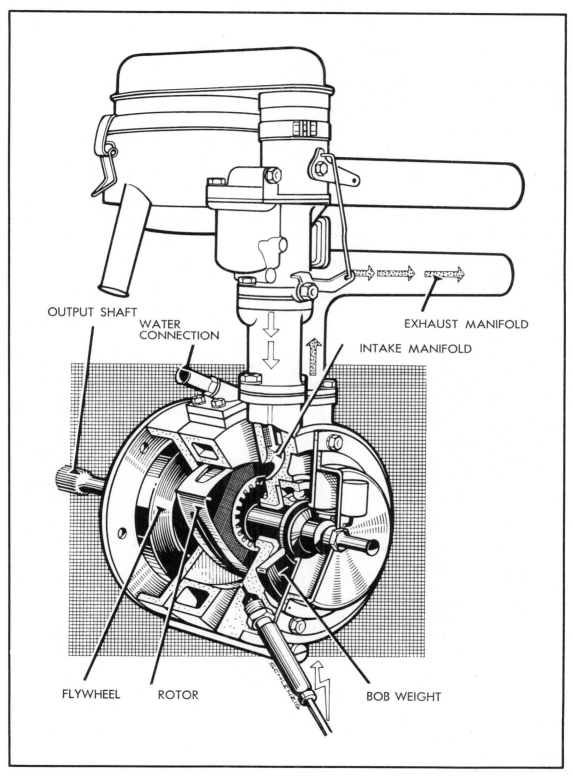

Fig. 17-2 Cross-Section through the NSU-Wankel Rotating Combustion Engine
Reveals Ingenious Yet Simple Design.

As these inventors and many others worked on rotating internal combustion engines two problems persisted: (1) the complex geometry involved, and (2) sealing the power or the pressures created by the burning, expanding gases. Sealing problems in the rotating engine were not so simple to solve as those presented by the round piston in a round cylinder as seen in a reciprocating engine.

Felix Wankel, a German engineer, was attracted to the rotating combustion engine, its problems, and its potential usefulness to man. For years, Felix Wankel worked in the area of developing better sealing techniques and gradually his expertise became known and he set up his own engineering and research laboratory. In 1951, Wankel made contact with Dr. Ing. Walter Froede of the NSU research department. NSU is an engine manufacturer in Germany. Their contacts eventually resulted in agreements that had the result of getting rotating combustion engines out of the laboratory and into industrial production. In 1958 the Curtiss-Wright Corporation, a company with long and broad experience in engine manufacture in the United States, became the licensee for the NSU/Wankel engine in North America. Curtiss-Wright and other licensees share technical progress with each other as these engines pass from the phase of limited production to that of mass production.

HOW THEY WORK

The rotating combustion engine, like the reciprocating engine, must: bring fuel into the engine, compress the fuel, ignite the fuel, allow the pressure of the burning fuel to move the "piston," and then allow the spent gases to exhaust themselves from the engine. But remember, this must be done in a rotating motion rather then a reciprocating motion.

The construction and configuration that makes this possible is a rotor (piston) that is a slightly rounded equilateral triangle rotating on an eccentric portion of the main drive shaft. The rotor apexes follow the shape of the epitrochoidal chamber. The geometry of the engine shows how the volume between any one of the rotor sides and the epitrochoidal housing will go through a cycle of enlarging and reduc-

ing as the rotor turns. The housing is provided with intake ports, spark plug, and exhaust ports, all at the proper locations to take advantage of the changing volume and provide for intake, compression, power and exhaust.

OPERATING CYCLE

A study of Fig. 17-3 will enable you to grasp the cycle of this engine. What you are looking at is equal to a three-cylinder piston engine since there are three sides to the rotor. We are following the action of side AC as it finishes its exhaust cycle and takes fuel into the chamber; side AB as it compresses the fuel, ignites the fuel and begins the power cycle; and side BC as it completes the power cycle and begins its exhaust cycle.

Positions 1, 2, 3, 4 show the uncovering of the intake port and the increasing volume of the chamber. Fuel mixture enters the chamber, being pushed in by atmospheric pressure. The engine has a conventional carburetor and the operating principle is the same as that applied to reciprocating engines.

Positions 5, 6, 7 show the changing geometry of the chamber during compression. The chamber becomes increasingly smaller, compressing the fuel mixture into a smaller and smaller space. Compression ratios can be designed to meet particular needs and they usually approximate those of similar size reciprocating engines; 8.6 to 1 is used in the NSU Spider automobile engine. At position 7, maximum compression is attained and a spark is triggered at the spark plug. The spark plug tip is recessed in a hole so the rotor can sweep by without hitting it.

Positions 8, 9, 10 show the expanding gases sweeping across the enlarging chamber and exerting pressure against the face of the rotor. This, of course, is equivalent to what we term the power stroke.

Positions 11, 12, and 1 comprise the exhaust cycle and the exhaust is scavenged from the engine.

The design of the engine permits the rotor to travel at one-third that of shaft speed. Speed ranges of 2000-17,000 r.p.m. have been reached.

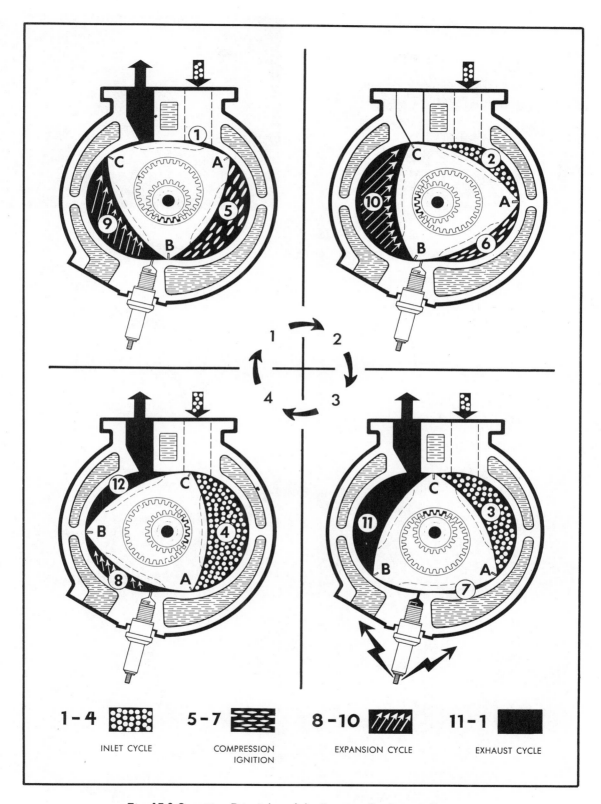

Fig. 17-3 Operating Principles of the Rotating Combustion Engine.

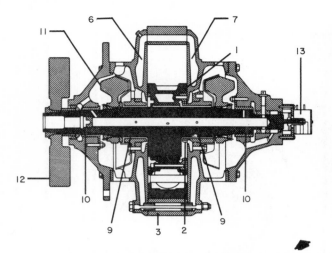

RC-60   LONGITUDINAL AND CROSS SECTIONS

1. Rotor With Internal Rotor Gear
2. Stationary Gear
3. Rotor Housing
4. Exhaust Port
5. Spark Plug
6. Side Housing -- Drive Side
7. Side Housing -- Anti-Drive Side
8. Intake Port
9. Main Bearing (Inner)
10. Main Bearing (Outer)
11. Balance Weight
12. Flywheel
13. Ignition Contact Maker

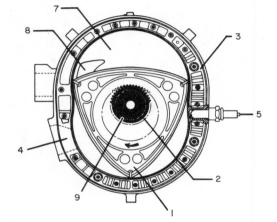

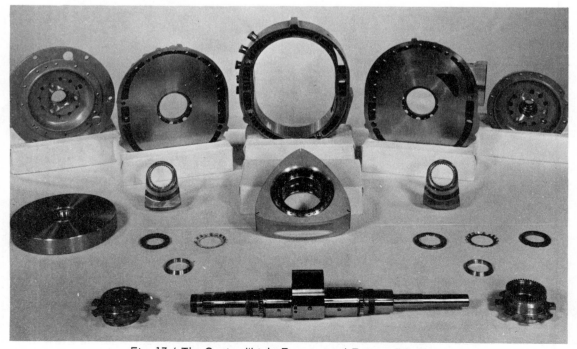

Fig. 17-4 The Curtiss-Wright Experimental Engine, RC-60.

COMPONENTS OF THE ROTATING COMBUSTION ENGINE

The rotor is a hollow casting of either aluminum or cast iron, Fig. 17-4. The three rotor flanks are exposed to a considerable concentration of heat; therefore, cooling oil is circulated through the rotor. Oil, under pressure from the oil pump, enters the rotor through drilled passages in the main shaft, Fig. 17-5, picks up heat, and then discharges back into the system. A large hole machined in the center of the rotor will accommodate the eccentric of the main shaft. A ring gear is secured to the rotor and meshes with a stationary gear on one of the end housings. These gears insure that the correct relationship is followed by the engine parts as the rotor travels the epitrochoidal path.

Devising effective seals for the rotor was accomplished only after exhaustive testing and development. The rotor must be sealed on the sides and also at the three rotor apexes. Fig. 17-7 illustrates the sealing methods now used.

The seals themselves are made of alloy cast iron that will not wear excessively itself nor will it cause excessive wear on the rotor housing. Light springs push the apex seals and side seals against the housing surfaces and provide an airtight seal. The apex seal may be traveling as a sliding velocity of up to 108 feet per second.

Lubrication in some form must be provided, just as piston rings must be lubricated. NSU/Wankel in Germany employs a separate oil supply delivered to the intake charge of fuel mixture. Curtiss-Wright lubricates through metering oil seal rings in the rotor itself. A rotating combustion engine will consume oil at about the same rate as a conventional reciprocating engine.

The forces that move the rotor are transmitted to the main shaft by means of an eccentric section of the shaft. The eccentric enables the rotor to follow the geometry of the housing. The main shaft is held and centered by the two end housings. One end will contain the various gear takeoffs for the engine such as the water pump, the oil pump, and the distributor shaft.

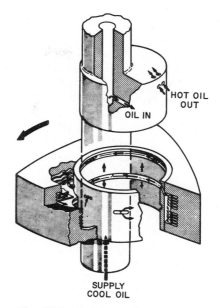

Fig. 17-5 Oil Flow through Main Shaft, Eccentric and Rotor.

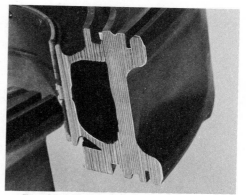

Fig. 17-6 Hollow Construction of Rotor.

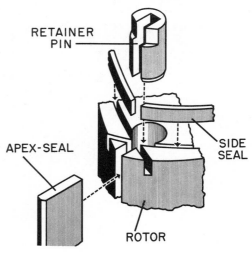

Fig. 17-7 Apex Seals and Side Seals for Rotor.

The cooling of the engine may be accomplished by either water or air. Cooling water is circulated around the main chamber housing and around the end units, Fig. 17-8. Special attention is given to the combustion area, Fig. 17-9. Combustion for each rotor flank always occurs at the same point and an abundance of cooling media must be present at this area. Air cooling is accomplished with finning, the cooling air being blown over the radiating fins, Fig. 17-11.

The performance of rotating combustion engines compares favorably to that of reciprocating engines, Fig. 17-10. Characteristics such as brake horsepower, thermal efficiency, and mechanical efficiency all indicate the adequacy of this engine. The torque curve for the engine is excellent with little drop at high speed. The flow of gases or volumetric efficiency of the engine is also excellent.

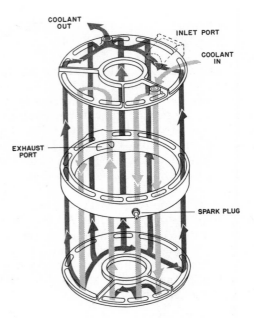

Fig. 17-8 Schematic of Housing Coolant Flow.

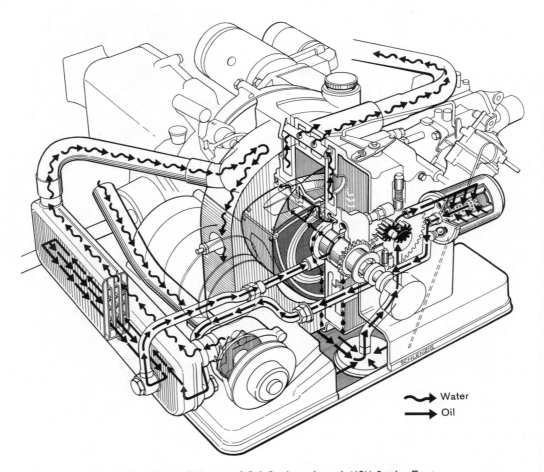

Fig. 17-9 Flow of Water and Oil Coolant through NSU Spider Engine.

RC6 TEST PERFORMANCE, FULL THROTTLE. SIDE INTAKE PORT

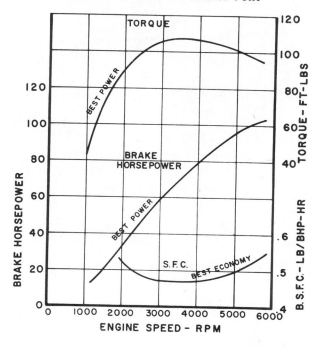

RC6 TEST PERFORMANCE, FULL THROTTLE, SIDE INTAKE PORT

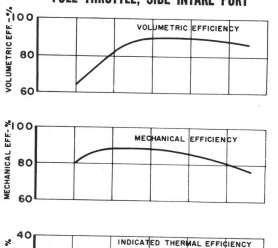

Fig. 17-10 Operating Characteristics of a Rotating Combustion Engine.

Fig. 17-11 Small Air-Cooled RC Engine.

Fig. 17-12 Internal Ribbing of End Housing. Note Side Intake Port and Cavities for Coolant Flow.

One significant change that Curtiss-Wright has made and incorporated into its version of the engine is the intake port, Fig. 17-12. The Curtiss-Wright engine has side intake ports that provide improved low-speed operation and

lessen fuel consumption, while the NSU-Wankel engine has peripheral intake ports located in the main housing. Both companies use peripheral exhaust ports.

The horsepower range of the engine is flexible. The Curtiss-Wright Corporation constructed one single rotor engine with a 782 brake horsepower rating, as well as auto size engines and low horsepower size engines. The horsepower of any basic engine can be boosted by merely adding rotors in succession. A twin rotor engine is equivalent to a six cylinder reciprocating engine. Three or four rotors and their chambers present no construction problems and they smooth out the operation even more.

Lightweight applications of the engine are being studied for use in aircraft. In automobile applications, the rotating engine is now in production. NSU/Wankel produces a single rotor 50 brake horsepower engine for use with its NSU/Spider automobile. The Toyo Kogyo Co. of Japan is also in limited production of an R-C powered auto. The future of the engine in the auto industry of the United States is promising but many technical and financial considerations will be weighed by the major manufacturers before decisions are made.

Fig. 17-13 A Rotating Combustion Engine Prototype, RC2-60, by Curtiss-Wright Corporation.

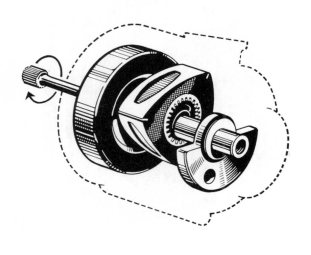

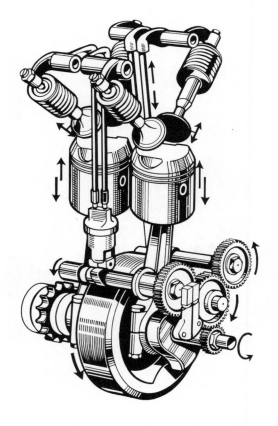

Fig. 17-14 Rotating Combustion Engines Require Fewer Parts, Are Lighter in Weight, Are More Economical to Manufacture than Reciprocating Engines.

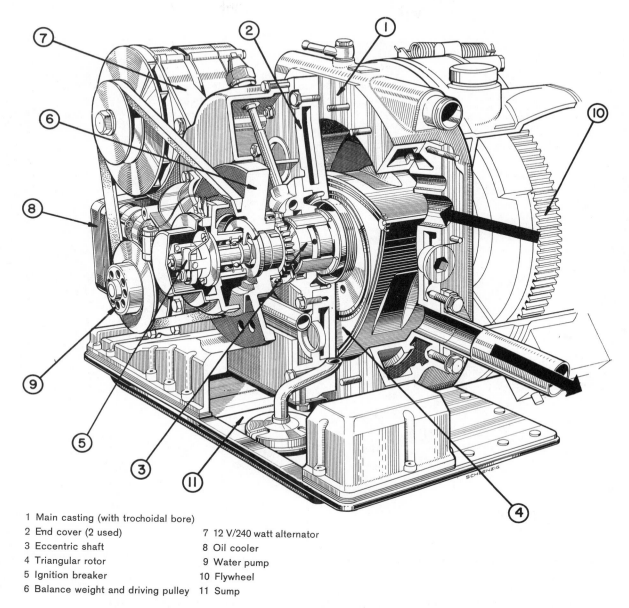

1 Main casting (with trochoidal bore)
2 End cover (2 used)   7 12 V/240 watt alternator
3 Eccentric shaft   8 Oil cooler
4 Triangular rotor   9 Water pump
5 Ignition breaker   10 Flywheel
6 Balance weight and driving pulley   11 Sump

Fig. 17-15 Cutaway View of the NSU-Wankel Spider Engine.

Smaller horsepower applications of the engine are also being developed and in some cases are in production: marine engines, portable pumps, generators, and so forth.

There is great potential for the rotating combustion engine; it can list many strong points and advantages.

1. Smooth operation with little vibration.

2. High horsepower-to-weight ratio.

3. Many different fuels can be used.

4. Wider speed range than that of reciprocating engines.

5. Excellent torque curve.

6. Economical to manufacture.

7. Fewer moving parts than reciprocating engines.

8. Can be produced in many different sizes.

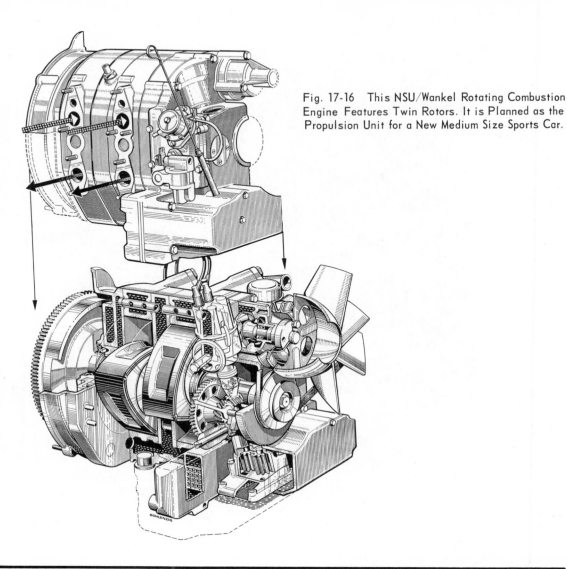

Fig. 17-16   This NSU/Wankel Rotating Combustion Engine Features Twin Rotors. It is Planned as the Propulsion Unit for a New Medium Size Sports Car.

GENERAL STUDY QUESTIONS

1. Explain the difference between the terms "rotating" and "reciprocating."

2. What two significant problems slowed the development of rotating combustion engines?

3. Who is credited with perfecting the design for the most widely used rotating combustion engine of today?

4. What name is given to the geometric shape of the rotor housing?

5. What four functions must be accomplished during a complete cycle of the rotating combustion engine?

6. Does the rotating combustion engine require a unique or specially designed carburetor, ignition system, water pump, or oil pump? Explain.

7. List the key moving parts on the rotating combustion engine.

8. List several advantages of the rotating combustion engine.

CLASS DISCUSSION TOPICS

- Discuss the role of technology in manufacturing and the development of the rotating combustion engine.

- Discuss the various problems that a "new" engine faces as it becomes an established and widely used product.

- Discuss the operating characteristics of the rotating combustion engine: Torque, Volumetric Efficiency, Mechanical Efficiency, Horsepower, etc.

- Discuss the operating principles of the rotating combustion engine.

Unit 18

STEAM ENGINES AND STEAM TURBINES

The steam engine is classed as a heat engine. The chemical energy of a fuel is released in the form of heat energy which the steam engine is designed to harness. Unlike most engines, the steam engine can be made to operate on almost anything that will burn - wood, coal, petroleum, or their many variations. The steam engine cares little what fuel is used, just as long as the water is boiled and steam is produced. Of course, the reason that steam engines can operate on many different fuels is because the fuel is not burned within the engine itself. Fuel is usually burned in a fire box under a boiler, outside the working parts of the engine. The engine is called an <u>external combustion</u> engine.

The energy path of a typical steam engine follows this route: (1) <u>Chemical energy</u> of the fuel is converted into, (2) <u>Heat energy</u> which boils the water to produce, (3) <u>Steam</u> which is confined in the boiler building up, (4) <u>High steam pressures</u> which when released and channeled, are able to, (5) <u>Move</u> the engine's <u>piston or turbine parts</u> which do useful work. It should be remembered that when water is converted into steam it expands, in fact, at the rate of 1600 to 1. It is not difficult, therefore, to see how steam pressures can be built up in a closed boiler.

It is almost impossible to credit any one man with the invention of the steam engine. Many scientists, engineers, and inventors contributed their ideas down through the years, each man building upon the accomplishments of his forerunner. Modern, powerful, efficient steam engines and turbines evolved from many crude, inefficient, and often unsuccessful efforts of earlier years.

If a starting point is to be fixed, it would be Hero's Aeolipile which was devised before the birth of Christ. The Aeolipile was basically a hollow sphere that was spun around by jets of steam coming from within the sphere itself. A "boiler" fed steam to the sphere. The action of the Aeolipile could be compared to that of water coming from a spinning lawn sprinkler. This "novelty" of the time began to open the door - the possibilities of steam power were becoming apparent.

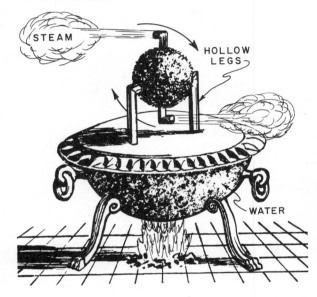

Fig. 18-1 Hero's Aeolipile - Steam Put to Work more than 2000 Years Ago.

Further progress was almost nonexistent until about 300 years ago when such men as Lord Worcester, who was followed by Captain Thomas Savery and Thomas Newcomen, began to work on engines to pump water from coal mines in England. Water, seeping into and flooding coal mines, had created a need for a good pumping device. Thomas Newcomen built steam engines that actually worked and several hundred were used throughout England's coal mines. His pump had a rocking action: a large "see-saw" beam with the pump on one end and the steam engine on the other. The action of the steam engine followed this pattern: (1) Coal burned under the boiler, boiled the water and produced steam. (2) Steam was let into the cylinder as the piston moved upward. (3) The steam valve was shut off. (4) Cold water was sprayed into the cylinder to condense the steam. (5) Atmospheric pressure pushed the piston

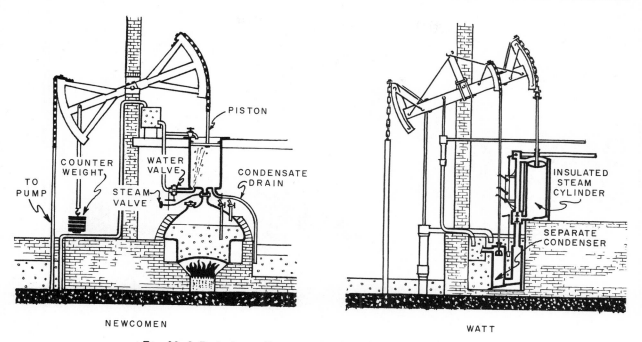

NEWCOMEN

WATT

Fig. 18-2 Early Steam Engines Produced by Newcomen and Watt.

back down the cylinder; the pump on the other end raised water out of the mine. The power delivered was actually produced when the steam was condensed and the atmospheric pressure, which was greater, pushed the piston down the cylinder. In a sense, Newcomen's engine was more an "atmospheric engine" than a "steam engine".

Newcomen's engines did consume an enormous amount of coal; they were very inefficient. As coal prices rose, the cost of operating these engines became a real problem.

The man with the answer to this problem was James Watt. He studied Newcomen's engine and detected some of its faults. He reasoned that the principle cause of inefficiency was that the cylinder was first heated by the steam, then cooled by the condensing water. This alternate heating and cooling wasted much power. His solution was to connect a separate tank for condensing the steam. With this condensing tank connected to the cylinder the cylinder could be kept hot all the time. Watt also devised a way to make the engine double-acting, that is, it would operate on both the push and pull strokes. The engines that Watt developed were much more efficient than Newcomen's and they were widely used in England.

Methods were developed to convert the up and down (reciprocating) motion of the engine into rotary motion. Engines became more versatile and they were soon found powering the wheels of many industries: paper mills, corn mills, cotton spinning mills, brewerys, and iron works. The steam engine became a "hand-in-glove" partner with the Industrial Revolution.

Watt's steam engines were heavy, and so they were built to remain in only one location. The answer to reducing the size and weight of steam engines lay in the use of high-pressure steam. Naturally, higher pressures required boilers that could withstand the pressures without bursting. Richard Trevithick, also an Englishman, successfully developed high-pressure boilers that operated at 60 pounds per square inch steam pressure.

Lighter, high-pressure steam engines were adaptable to vehicles. The locomotive lent itself best to the use of steam. Trevithick built a steam locomotive and, though it was not entirely successful, it did work. George Stephenson was the engineer who brought together all of the best steam propulsion ideas to build a practical steam locomotive. His "Rocket" sparked the rapid development of the railroad industry.

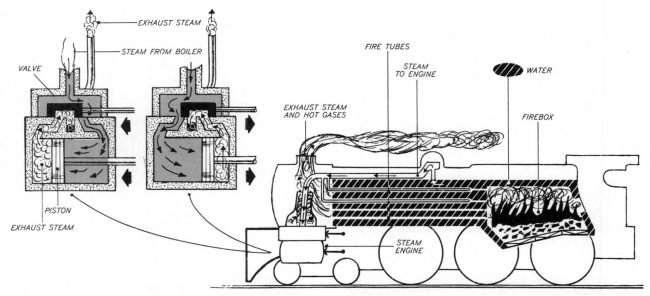

Fig. 18-3   A Typical Steam Locomotive.

The modern day steam engine has changed little in its basic components as it has developed. Two factors have contributed to the steam engine's increasing efficiency, (1) higher steam pressure and (2) higher steam temperature. A small amount of steam placed under conditions of high pressure and high temperature is capable of producing considerable power as it expands, pushing the piston down the cylinder. Steam locomotives that produce as much as 6500 hp. have been built. Steam pressures on locomotives have increased to the range of 300 lbs. per square inch (p.s.i.).

Even though the steam locomotive has been replaced by the diesel-electric locomotive, it will be worthwhile to consider what takes place within a typical railroad locomotive.

Fuel is burned in the firebox and the hot gases of combustion are drawn through the boiler. The fire tube boiler consists of many steel pipes surrounded by water. As the hot gases sweep through the pipes, some of the heat (the more the better) is transferred to the water in the boiler. The water boils and the steam rises to the top of the boiler where it is collected in the dome. The steam next travels through superheaters where its temperature is increased. The superheated steam is introduced into the steam chest and is ready to deliver its energy. The sliding valve within

the steam chest is designed to deliver steam to one side of the piston. When the piston has completed its stroke, the sliding valve admits steam to the other side of the piston. Power is produced on both strokes of the piston. The steam is exhausted up the exhaust stack of the locomotive, helping to create a good draft through the engine.

Some steam engines are compound engines, that is, the exhaust steam from one cylinder is lead into a second (or third) cylinder to expand further, extracting more of the steam's energy. On compound engines, each succeeding cylinder must be larger. Most steam locomotives are not of the compound type. Rather, they have two pistons, one on either side of the locomotive.

Also, some steam engines are equipped with condensers to conserve the boiler's water supply. The steam engines on ships need condensers since salt water cannot be used in a boiler without building up large deposits in the boiler. It would not be practical to carry enough fresh water to replace the water that is boiled away; the used steam must be condensed back into water and returned to the boiler. The railroad locomotive does not need a condenser, here it is more practical to stop periodically and take on more water for the boiler replacing the water that is lost as steam.

STEAM TURBINE

The idea of a steam turbine can be traced back to 1628 when Giovanni Branca designed a steam turbine. Though his ideas are on record, it is not known whether or not his turbine was actually built. From that time, little was done with the turbine until some years after Watt's reciprocating steam engine was developed and in wide use. At this time, engineers and scientists again began to think in terms of steam turbines. They reasoned that much power was being wasted in driving the reciprocating engine's many cranks, connecting rods, and in the start-stop, back-and-forth motion of the piston.

Engineers of that day envisioned a steam turbine with smooth, continuous power, like a windmill or water wheel. The first to build a successful turbine was a Swedish engineer by the name of Carl DeLaval. His turbine, built in 1882, consisted of a wheel with vanes mounted on the rim; four nozzles directed jets of steam against the vanes to drive the turbine. The action was much the same as wind driving a windmill. The kinetic force of the steam struck the blades, giving up a portion of its energy to revolve the turbine. This type of turbine is called an impulse turbine.

DeLaval designed a nozzle that would take advantage of the behavior of steam as it is released from a boiler. Naturally, the steam expands, but it also increases in speed or velocity as it is released. The DeLaval nozzle helped to increase the velocity and kinetic force of the steam.

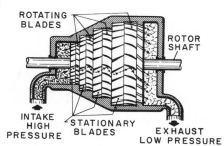

Fig. 18-4 A Simplified Steam Turbine.

In order to be most efficient, the turbine must rotate at about one-half the velocity of the incoming steam. This means that the turbine rotor speeds in a range of 10,000 r.p.m. to 30,000 r.p.m. Such speeds create problems in themselves since the speed must be geared way down to operate machinery. The characteristics of DeLaval type turbines make them best suited for installations that require a relatively small amount of power.

C. G. Curtis, an American inventor, used DeLaval's ideas to produce a more powerful impulse turbine. Curtis constructed his turbine with a series of revolving blades and stationary blades. In today's Curtis turbine, the steam hits the first rotor vanes, imparting some motion to the rotor as the steam changes its direction of flow (a V-like, glancing travel). The succeeding set of stationary vanes changes the steam direction back again and the steam strikes a second set of rotor vanes, giving up more of its kinetic energy. The next set of stationary blades reverses the steam direction again and the process continues as the steam travels through the turbine.

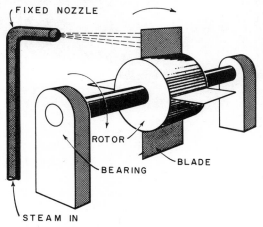

Fig. 18-5 Simplified Impulse Turbine.

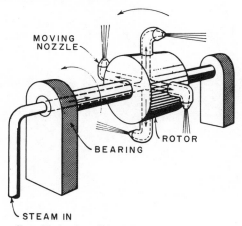

Fig. 18-6 Simplified Reaction Turbine.

Often turbines are divided into two or more pressure sections to gain increased efficiency. Each diminishing pressure section has larger turbine rotors to accommodate the expanded steam. These turbines are the compound impulse type.

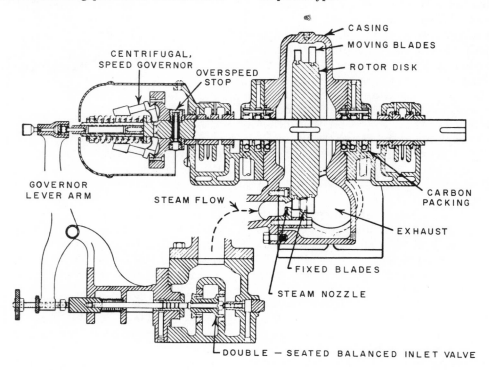

Fig. 18-7   Cross-sectional View of a Small Impulse Turbine.

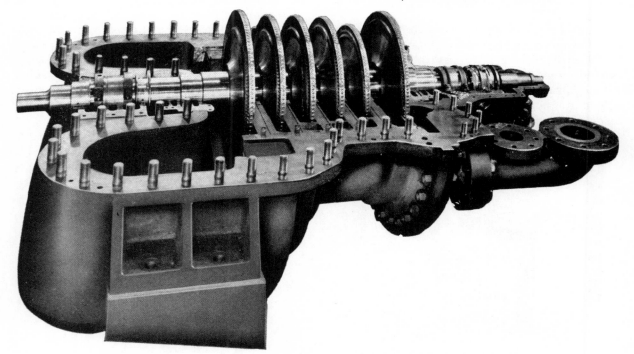

Fig. 18-8   A Six Stage Impulse Turbine With the Top Half of the Casing Removed.

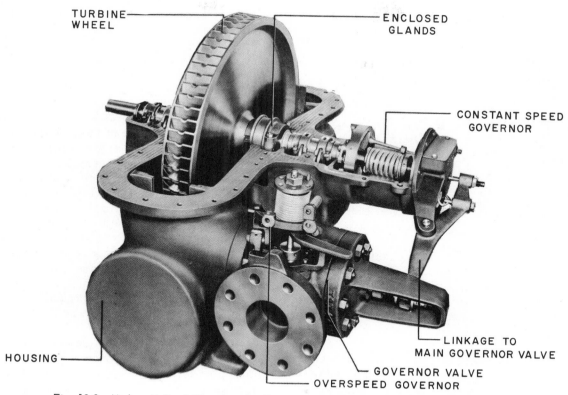

TURBINE WHEEL

ENCLOSED GLANDS

CONSTANT SPEED GOVERNOR

LINKAGE TO MAIN GOVERNOR VALVE

GOVERNOR VALVE

OVERSPEED GOVERNOR

HOUSING

Fig. 18-9   Modern Helical Flow Impulse Turbine with the Top Half of the Casing Removed.

Fig. 18-10   Action of the Steam in a Solid Wheel Machine.

Another type of impulse turbine in use today is the helical flow impulse turbine. In this turbine the steam follows a spiralling path as it delivers its kinetic energy to the rotor.

Our most widely used and most important turbine is the reaction turbine. This turbine was invented by C. A. Parsons of Ireland. His turbine was similar to Curtis's in that it had alternate rotating blades and fixed blades but the design of the blades themselves was different. Instead of having a uniform cross-sectional path to follow, the steam followed a converging cross-sectional path. As a result, the steam was accelerated. When it left the turbine blade, the steam produced the reaction of adding push to the rotor. The principle is identical to that of a jet or rocket engine. For every action there is an equal but opposite reaction; the action of the accelerated steam causes the reaction of pushing the rotor forward. Steam that enters a large turbine at 75 m.p.h. under high pressure will leave the turbine at 300 m.p.h. under low pressure. The steam can pass through a thirteen foot turbine in 1/16 of a second.

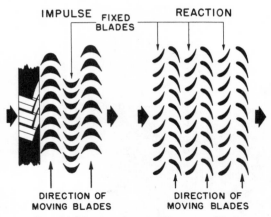

Fig. 18-11 Impulse and Reaction Blading.

In actual practice the reaction turbine does not derive 100% of its power from the reaction principle. About half of its power comes from impulse - half from reaction. Most modern steam turbines are, therefore, a blend of impulse and reaction.

Turbine efficiency is increased by adding a condenser to the low-pressure end of the turbine. The condenser produces very low pressures at the end of the turbine which enables the initial steam pressure to be reduced.

The general pattern in turbine construction is to reduce the steam pressure in many stages for slow-speed operation or reduce the steam pressure in few stages for high-speed operation. A high-speed turbine used for driving a generator may reduce the steam pressure in 50 steps while a low-speed turbine for propelling a ship may reduce the steam pressure in 200 steps.

The primary application of steam turbines is in the field of generating electricity. The high-speed turbine is well suited to this purpose since generators also operate best at high speed. The steam turbine accounts for two-thirds of the total electrical output in the United States. Most often the electrical generating plant will be located very near a river which provides a constant source of fresh water and economical barge transportation of the coal that is used as fuel. The coal is pulverized and blown into the fire chamber of the boiler where it literally explodes as its chemical energy is changed into heat energy. Steam is produced in the many water tubes in the boiler and the steam is of course led to the turbines which drive the generators. Steam turbine electrical generating plants have a high efficiency, 20 to 25 percent.

Modern turbines are designed to meet a large variety of applications. Different speed horsepower requirements require the turbine manufacturers to develop many turbine styles although the basic operating principles remain the same. Many modern turbines use steam under as much as 1350 p.s.i. pressure at 950°F. Pressure and heat of this nature were possible only after metallurgists were able to develop alloys that could stand up. Careful machining and workmanship go into every turbine, making it an efficient prime mover of watch-like precision.

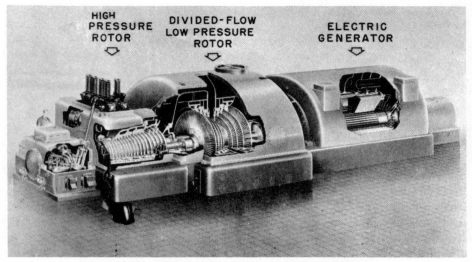

Fig. 18-12 A Typical Steam Turbine Generating Plant.

GENERAL STUDY QUESTIONS

1. Explain why a steam engine is classed as an external combustion engine.

2. Briefly trace the energy path within a steam engine or steam turbine.

3. What were the first steam engines used for?

4. Why is Thomas Newcomen's steam engine sometimes referred to as an "atmospheric engine"?

5. In what way did James Watt improve the steam engine?

6. What improvements made the reciprocating steam engine adaptable to locomotives?

7. Explain why some steam engines need a condenser and some steam engines do not.

8. Explain the action of the steam in an impulse turbine.

9. Trace the path of the steam through a Curtis-type turbine.

10. In what way is a reaction turbine different from an impulse turbine?

11. Are modern turbines pure impulse or pure reaction? Explain.

12. What is the most important use of steam turbines today?

13. What is the efficiency of a modern steam turbine?

CLASS DISCUSSION TOPICS

● Discuss the expansion of water as it turns into steam and how this action can be converted into power.

● Discuss the advantages of increasing steam pressure and steam temperature.

● Discuss the decline and fall of the steam locomotive in the railroad industry.

● Discuss the advantages of a steam turbine over a reciprocating steam engine.

● Discuss how it is possible for steam to accelerate as it passes through a turbine.

● Discuss impulse and reaction turbines.

● Discuss the steam turbine electrical generating plant.

Section IV
ELECTRICAL ENERGY

Unit 19

PRINCIPLES OF ELECTRICITY

If a person is to appreciate electricity, he should try to place himself at a time in history before electricity was commonly used. There were no electric lights, radios, television sets, telegraphs, telephones, or electric motors. There were no power household appliances: washers, driers, vacuum cleaners, toasters, electric coffee pots, or hair driers. The machines of industry were powered by large, inefficient steam engines. The pace of life was slower before the era of electrical technology; however, both men and women worked harder than their modern counterparts do. Aided by electricity, our work load has lightened.

Communication devices such as radio, telephone, telegraph, and television have speeded communications and helped to "shrink" our world. Messages can be transmitted at the speed of light. While the world has, in a sense, been made smaller, our understandings of other peoples have been broadened. Modern communication places ideas and pictures before us from all parts of the world. Today, we are made aware of community, state, national, and world affairs and have an opportunity to expand our understanding and knowledge.

The "romantic" era of candle light and horse drawn carriages is gone. But remember, it was also the era of long, hard working hours and low productivity. Electricity in industry, farm and home has enabled all persons to enjoy a richer, fuller life.

Electron theory and the characteristics of the atom were discussed in Unit 7. You will recall that a source of electricity can be mechanical, chemical, or static. In Unit 7 the mechanical source of electricity was discussed. This unit, therefore, will cover the remaining two sources, static electricity and the electro-chemical production of electrical power.

STATIC ELECTRICITY

It is possible for some substances to build up an excess of electrons and for others to have a deficiency of electrons. A body with an excess of electrons is said to be <u>negatively charged</u>. A body with a deficiency of electrons is said to be <u>positively charged</u>.

Just as electrons and protons are attracted to each other, <u>unlike</u> <u>charges</u> <u>attract</u> <u>each other</u>; negative charges and positive charges attract each other. The converse is also true, <u>like</u> <u>charges</u> <u>repel</u> <u>each</u> <u>other</u>; negative repels negative and positive repels positive. The Greeks observed these electrical properties 2500 years ago though they did not fully understand them. Their observations pertained to what is called <u>static</u> <u>electricity</u>.

Substances differ in their ability to lose or gain electrons. For example, if a hard rubber rod is rubbed with fur, the rod becomes negatively charged. Electrons from the fur are brought in contact with the rubber. Some leave their atoms, building up an excess of electrons on the hard rubber rod. Another example is a glass rod that is rubbed with silk. In this case, however, electrons on the glass break free and are transferred to the silk. The rod is short of electrons, positively charged.

The action of static electricity can best be seen through the use of an <u>electroscope</u>. Basically, the electroscope is a glass container with a metal rod that passes part way through its stopper. Attached to the lower end of the rod are two metal leaves. When the electroscope is neutral, the metal leaves hang down. However, when a negatively charged rubber rod touches the electroscope, many electrons flow into the rod and leaves. The leaves fly apart or diverge because they are both now negatively charged: like charges repel.

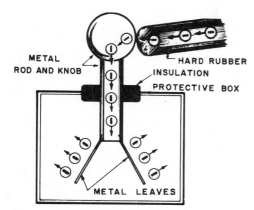

Fig. 19-1 A Negatively Charged Electroscope.
Electrons are Received from the Negatively
Charged Rubber Rod.

If a positively charged glass rod is brought in contact with a neutral electroscope, electrons flow from the electroscope onto the glass rod. The metal leaves are both short of electrons; they are positively charged and they fly apart because like charges repel. Charging by either of these methods is called charging by <u>contact</u>.

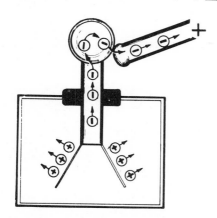

Fig. 19-2 A Positively Charged Electroscope.
Electrons are Taken from the Electroscope by
the Glass Rod.

Charging can also be done by <u>induction</u>. Suppose that a negatively charged hard rubber rod is brought close to an electroscope. Electrons in the electroscope are repelled. Now, if a person touches the electroscope providing an escape path for the electrons, some electrons will flow onto the person's finger. When the finger is removed, the electroscope is short of electrons and positively charged. The leaves fly apart and remain apart even after the rubber rod is removed from the area. Positive charges can also be induced onto the electroscope by using a positively charged glass rod.

The electroscope can store an electrical charge. Another device, the Leyden jar, can also store an electrical charge. For modern applications, a condenser is used to store electrical charges.

Electrons want to travel from a negative area to a more positive area. Unlike charges attract. Electrons will move until both areas are neutral. Static charges want to neutralize themselves; they can do this if (1) they are connected with each other, perhaps by a wire, (2) they are in contact with each other, (3) they are brought close enough together for the electrons to jump from the negative body to the positive body: this can be seen as an arc or spark. Electrons tend to flow when there is a difference in charge or electrical potential.

ELECTRONS IN MOTION

Electric current may be defined as a movement of electrons within a conductor toward a more positive body. The basic unit of electron measurement is the <u>coulomb</u>. A coulomb is 6,250,000,000,000,000,000 electrons. Another basic measurement is the <u>ampere</u>. If one coulomb of free electrons passes a given point in one second, one ampere of electrical current is flowing.

Why do electrons flow? They flow because there is a difference in electrical potential, and this difference is called <u>voltage</u>. Further, electrons flow from the negative to the positive potential.

Another factor, besides voltage, that enters into the flow of electrons is <u>resistance</u> measured in <u>ohms</u>. Some materials are good conductors and offer little resistance to electron flow. Large diameter wires offer less resistance to electron flow than small diameter wires. The length of the wire also affects resistance; the longer the wire, the greater the resistance.

Ampere - the unit measurement of the rate of electron flow.

Volt - the unit measurement of electrical potential or pressure.

Ohm - the unit measurement of the resistance a substance offers to current.

The three factors, current, voltage, and resistance, are inseparable; each is related to the other. Their relationship is expressed mathematically as Ohm's Law:

$$I = \frac{E}{R} \qquad R = \frac{E}{I} \qquad E = IR$$

I = Current intensity or amperes

E = Voltage or electromotive force

R = Resistance stated in ohms

Three simple diagrams below show how Ohm's Law can be used to calculate electrical problems.

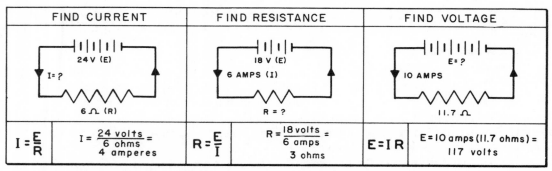

Fig. 19-3 Applications of Ohm's Law.

ELECTRICAL CIRCUITS

In discussing electrical circuits, we shall assume that we have direct current (d.c.), a continuous flow of electrons in one direction. Typical producers of direct current are batteries and d.c. generators. Current will flow from a battery, for example, as long as there is a complete path for it to follow, a complete circuit. A switch is usually placed in the circuit to control electron flow. When the switch is closed, electrons flow, the light is on etc. When the switch is opened, current stops. When the switch is open, the electrical potential is still present but there is no way for it to be satisfied. The circuit is not complete. A complete circuit is necessary for electron flow.

There are two basic types of electrical circuits: (1) series and (2) parallel. If all the resistances in the circuit are in one path, the circuit is a series circuit. Current must pass through all of the resistances; therefore, the total resistance of a series circuit is the sum of the individual resistances. A voltage drop can be measured across each resistance. The total voltage is equal to the sum of the separate voltages. In Fig. 16-4 if one lamp burns out, the circuit is broken and all the lamps go out. This is one disadvantage of series wiring.

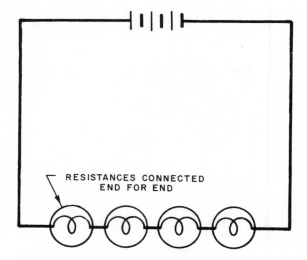

Fig. 19-4 Series Circuit.

Voltage is divided among the resistances; current is the same through all resistances; resistance is the sum of the separate resistances.

If the resistances in a circuit are connected side by side, each having one end to the power source, the circuit is a parallel circuit. Current divides itself to flow through the available paths. The amount of current that flows through any part depends on the resistance of that part. The resistance is divided up so that the total parallel resistance is less than the smallest single resistance in the circuit. The voltage across each resistance is the same, the full 117 volts on normal household installations.

Voltage is the same across all resistances; current is divided among the resistances; total resistance is less than that of the smallest resistance.

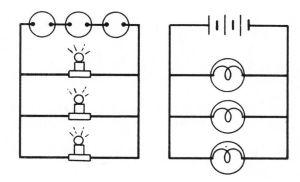

Fig. 19-5 Parallel Circuit.

MAGNETISM

The ability of certain metals to attract iron or steel was seen in Unit 7. These metals, called magnets, can also attract nickel, cobalt, chromium, and magnesium but to a lesser degree. Although the first practical use for the magnet was the compass, this is not our main reason for studying magnetism. Today, magnetism has achieved greater importance because most of the electricity that is produced originates with the principles of magnetism.

You will recall that in explaining the phenomenon of magnetism it was reasoned that atoms of iron and a few other substances have the special ability of lining themselves up in a pattern whereby the magnetic qualities of each atom complements those of surrounding atoms. Atoms work together, combining their minute magnetic fields to produce a noticeable magnetic field. In nonmagnetic substances atoms are arranged in a random manner and, therefore, they do not exhibit magnetic properties.

Substances that are magnetic have a north-seeking pole and a south-seeking pole, corresponding to the earth's magnetic poles. In effect, each magnet has a magnetic field around it just as the earth does. As a reminder, the like poles of two magnets repel each other while the unlike poles attract each other.

The magnetic field is strongest at the ends of the magnet. In the center of the magnet the strength is negligible. A magnetic field can be seen by placing a magnet under a piece of paper and then sprinkling iron filings onto the paper. The iron filings line up to show the magnetic field and its concentration.

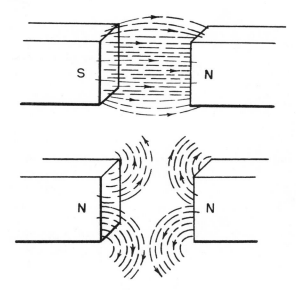

Fig. 19-6 Unlike Poles Attract Each Other; Like Poles Repeal Each Other.

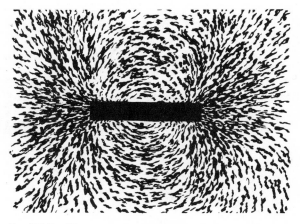

Fig. 19-7 Illustrating a Magnetic Field by the Use of Iron Filings.

ELECTROMAGNETISM

When electric current flows through a wire, a magnetic field is built up around the wire itself. This can be seen using a cardboard, iron filings, and a current-carrying wire. With the wire passed through the cardboard and the current flowing, iron filings are sprinkled onto the cardboard. They can be seen arranging themselves in a magnetic field. The magnetic lines of force are referred to as flux. Just as in a natural magnet, the field is strongest near the wire and diminishes as the distance from the wire increases.

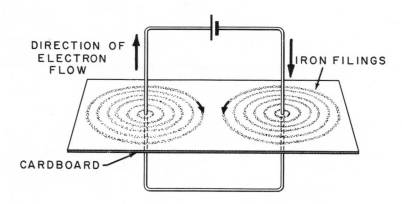

Fig. 19-8   Magnetic Field Around a Wire that is Carrying Electric Current.

Flux around a wire does have direction. Flux direction is determined by the direction of electron flow within the wire. As shown in Fig. 19-9, the North Pole of the compass needle indicates the direction of flux or magnetic field around the wire. The dot in the center of the wire on the left indicates the point of the current-direction arrow coming toward the observer; the X at the right represents the tail of the current arrow pointing away from the observer. If the direction of electron flow within the wire is reversed, the compass needles will reverse themselves, indicating a change in flux direction.

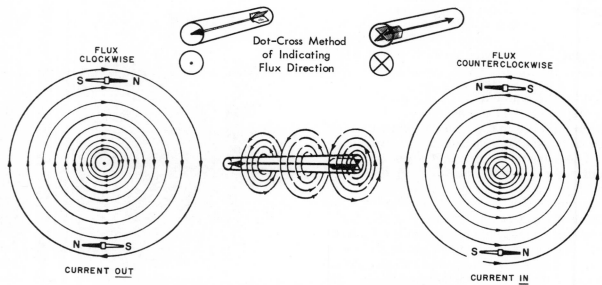

Fig. 19-9   Compasses Indicate the Direction of Flux Around a Wire.

The Left-Hand Rule will enable a person to determine either flux or current direction, providing one of the two is known. If current direction is known, point the left thumb in the direction of flow. When your fingers grasp the wire, they point in the flux direction. If the flux direction is known, wrap the fingers of the left hand around the wire in the direction of the flux. The left thumb will point in the direction of current.

Fig. 19-10   The Left-Hand Rule for Determining Flux Direction or Direction of Electron Flow.

In Unit 7 it was seen that when a current-carrying wire is made into a coil, the whole coil becomes a magnet; a single magnetic field builds up around the coil. You might say that the magnetic flux around the turns of wire combines to form one large magnetic field. The coil is said to be electromagnetic.

The magnetic field of the electromagnet has a North Pole and a South Pole just as a natural magnet has. The like poles repel and the unlike poles attract. Electrically speaking, the ends of the coils where the currents are parallel in the same direction are attracted to each other; the ends of the coils where the currents are parallel in the opposite direction repel each other.

On the electromagnet, a soft iron core is used to strengthen the magnetic field. Electromagnets and other magnetic devices can be used for lifting iron and steel, magnetic separators, relays, bells, solenoids, sound producers, and many other items.

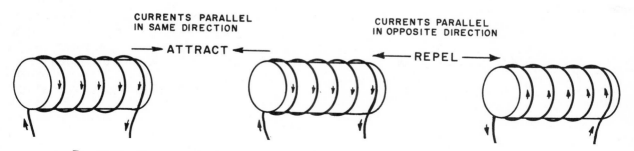

Fig. 19-11   Electromagnets Attract or Repeal Each Other According to the Direction of Current in the Coil.

ELECTROCHEMICAL SOURCES OF ELECTRICAL ENERGY

Dry cells and automobile storage batteries are perhaps the two most common examples of the conversion of chemical energy into electrical energy. There are certain differences between the two which class them as primary cells and secondary cells. First of all, the primary cell, such as a dry cell, is a single complete unit whereas the secondary cell, such as a storage battery, is a group of cells. Another point of classification is that primary cells cannot be recharged once the chemical reaction is complete; a secondary cell, on the other hand, can be recharged and used for a long period of time.

The dry cell is essentially made of two electrodes and an electrolyte. Its operation depends on a difference in the electrical potential of the two electrodes. Equally important, it depends on ionization for its operation. An ion is an electrically-charged atom or group of atoms. They may have lost electrons or gained electrons. Many substances are capable of being ionized.

The dry cell, which really is not dry at all, is made of a zinc outside shell or container; this is one electrode. The container is filled with a solution of ammonium chloride (NH_4Cl) dissolved in water; this is the electrolyte. Extended down the center of the cell is a carbon rod, another electrode. The top of the cell is sealed off, preventing evaporation of the pasty electrolyte. Free electrons in the zinc travel through the wire to the carbon electrode. These electrons are picked off by ammonium ions in the electrolyte. The zinc case is the negative terminal and the carbon rod is the positive terminal. There is a potential difference of 1 1/2 volts between the two electrodes.

The storage battery converts chemical energy into electrical energy. The battery case contains alternate plates of lead peroxide and spongy lead, both active but chemically quite different. Separators prevent the plates from touching each other. The battery is filled with a solution of sulfuric acid and water. Electrically speaking, the spongy lead plates are cathodes, or negative section, lead peroxide plates are the anodes or the positive section, sulfuric acid and water is the electrolyte. There is a voltage difference between the two groups of plates and when the circuit is closed (auto started) current discharges from the battery. When the circuit is closed the electrolyte attacks the lead plates; when the current

is in motion, the battery is said to be discharging. The lead plates start to become coated with lead sulfide; they become more alike. The sulfuric acid breaks down forming the lead sulfide and water. The water further dilutes the electrolyte. When the battery is discharged the plates are very similar and there is little voltage difference between the two plates. Voltage depends on the chemical difference between the two active materials.

Since the solution of sulphuric acid and water gets weaker or less dense as the battery discharges, the condition of the battery can be determined by checking the specific gravity of the solution with a battery hydrometer. The solution in a fully charged battery will have a specific gravity of 1.300; in a completely discharged battery, the solution will have a specific gravity of 1.120. Remember, the specific gravity of water is 1.000.

Fortunately the action of the battery can be reversed, recharging the battery without damaging it. The automobile's generator serves this purpose. The battery will be recharged when (1) the voltage produced by the generator is greater than battery voltage and also (2) the current being drawn from the battery is not greater than that the generator can put in. Thus, the chemical action will be reversed, returning the battery to its charged condition.

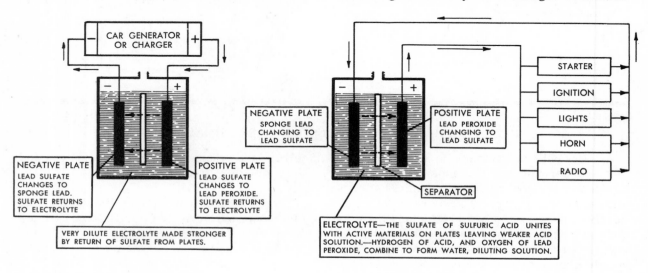

(A)   DURING CHARGE   (B)   DURING DISCHARGE

Fig. 19-12 Electrochemical Action in a Battery.

The lead storage battery is used in automobiles as a portable source of electrical energy. Six-volt batteries have three cells connected in series and twelve-volt batteries have six cells connected in series. The electrodes of the battery are alternate plates of lead and lead dioxide. The lead plate is the negative electrode and the lead dioxide plate is the positive electrode. The electrolyte is a solution of sulfuric acid and water. Electrons flow from the negative lead plate to the positive lead dioxide plate. Energy comes from the tendency of lead atoms to give up two electrons to each lead ion at the positive plate. Ionization within the electrolyte is also a part of the action. This chemical action can take place only where the plates are in contact with the electrolyte. To produce a large current, the plates are made so that a large surface area is in contact with the solution. Separators of wood, glass fibers or similar porous material keep the plates from touching each other.

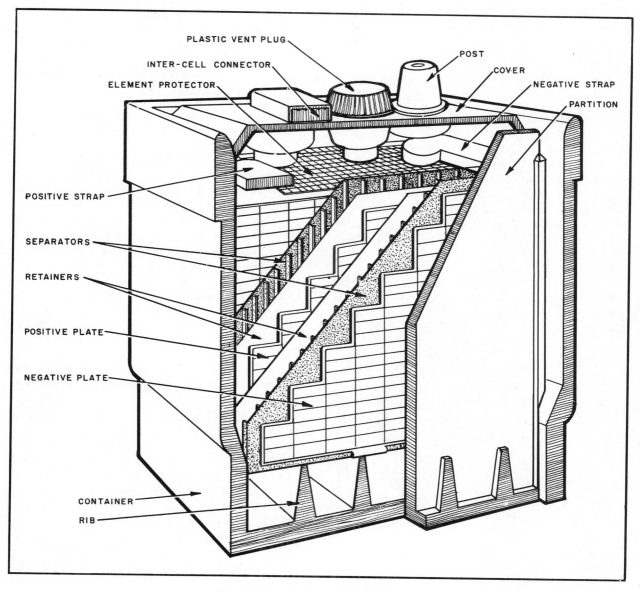

Fig. 19-13   The Lead Storage Battery.

INDUCTION

Whenever magnetic lines of force cut a wire, electric current is induced to flow in the wire. It should be pointed out that there must be a changing flux in order to induce electron flow. If the wire is held stationary in a fixed magnetic field there is no electron flow. Electron flow occurs when: (1) the wire moves relative to the magnetic field, (2) the magnetic field moves relative to the wire.

Either an increasing (expanding) magnetic field or a decreasing (collapsing) magnetic field can cause electron flow. More current can be induced by: (1) increasing the speed that magnetic lines of force are cut, (2) increasing the strength of the magnetic field.

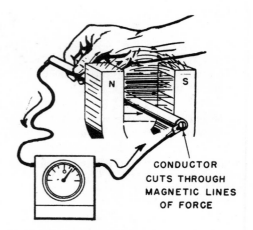

CONDUCTOR
CUTS THROUGH
MAGNETIC LINES
OF FORCE

Fig. 19-14 Producing an Induced Current by Moving a Wire Through a Magnetic Field.

D. C. and A. C. ELECTRICAL GENERATORS

Electrical generators are used to convert mechanical energy into electrical energy. Energy of free falling water directed against a turbine blade or the energy of hot gases directed against a turbine blade produces mechanical energy. This mechanical energy is used to produce electricity. The real key to the production or generation of electricity is the ability of a magnetic field to cause electron flow within a wire.

In a generator, the magnetic field is of constant intensity and the magnetic field is stationary. The coil of wire is moved within the magnetic field. A turbine is usually used to revolve the coil of wire: mechanical energy is used to produce electrical energy.

The magnetic field is normally produced by electromagnets that are designed to produce a strong magnetic field. The coil of wire, the armature, rotates inside the magnetic field. Actually, the armature is made of a shaft that has many iron laminations fastened to it. The laminations strengthen the magnetic action. Conducting wire is wound around the laminations. The two ends of the wire are connected to a slotted slip ring mounted on the armature. Two brushes, one on either side of the slip ring or commutator, provide the necessary path for electron flow.

For the sake of clarity the drawings and explanation of the d.c. generator have been simplified. In position (1) there is no electromotive force produced since the wire (armature) is moving parallel to the field's magnetic lines of force. However, as soon as the armature begins to move toward position (2), it begins to cut more and more magnetic lines of force. At position (2) the maximum number of magnetic lines of force are being cut. Induced current has gradually built up. Electrons flow toward the right-hand brush. As the armature continues on toward position (3), induced current continues to flow in the same direction but at a diminishing rate. At position (3), there is no electromotive force generated. Also, the

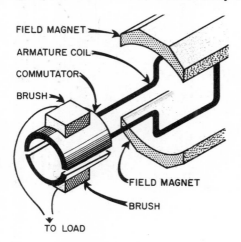

FIELD MAGNET
ARMATURE COIL
COMMUTATOR
BRUSH
FIELD MAGNET
BRUSH
TO LOAD

Fig. 19-15 A Simplified D-C Generator.

right-hand half of the ring will now touch the left-hand brush and vice versa. Now the armature moves up to position (4) again cutting lines of force. The direction of armature coil movement has been changed in relation to the magnetic field and the direction of electron flow has been reversed. However, since the commutator segments are touching opposite brushes, the effect is that the current continues in the same direction.

As the armature continues its rotation, it produces a constant stream of electrons flowing in one direction. Electrons flowing in one unchanging direction are said to be a direct current.

In actual construction, the d.c. generator is much more complex than these simplified diagrams indicate. In order to smooth out the operation of the generator and to produce a steady electromotive force, several coils of wire are wound around the rotating iron core. This means, of course, that the construction of the commutator must also be much more complex.

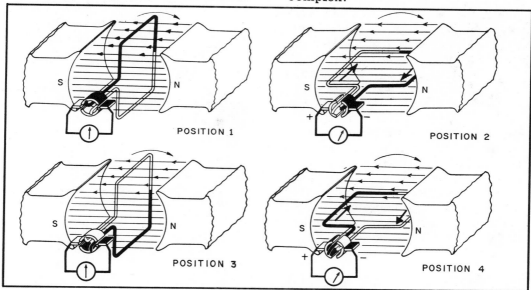

Fig. 19-16   Direct Current Generator Rotation.

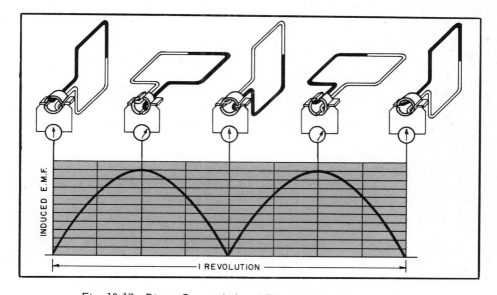

Fig. 19-17   Direct Current Induced Electromotive Force Graph.

The alternating-current generator is very similar in its construction and operation to the direct-current generator. However, the current produced by the a.c. generator changes its direction with every revolution of the armature. Instead of having a split ring commutator, two complete rings are mounted side-by-side on the armature shaft and connected to the armature coil. A brush rides against each ring, one brush always in contact with the same ring. In Fig. 17-18 the basic action shown is the same as that of a d.c. generator. At position (1) of the armature there is no electromotive force produced but as the armature moves on toward position (2) the induced electron flow increases. Current reaches its maximum at position (2), diminishing to zero as it arrives at position (3). The direction of electron flow reverses when the armature starts moving toward position (4). The induced electron flow reaches its maximum at position (4), diminishing to zero as it arrives back at position (1). The slip ring commutator enables the generator to produce a flow of electrons that changes its direction with every armature revolution or cycle. Ordinary alternating current is supplied at 60 cycles. This means that there are 60 complete cycles or revolutions (120 changes of current direction) every second.

Both alternating current and direct current are in common use. Both are good for the operation of electric light, heating units, or electric motors (a.c. or d.c.). Direct current is needed for electroplating, electrolysis, or the charging of storage batteries. However, the use of alternating current is more widespread because it can be transmitted over long distances more easily than direct current can. The voltage of alternating current can be stepped up or stepped down by the use of transformers. Generally, the voltage from the power station is stepped up to be transmitted over a long distance and then stepped down near the point where it is to be used. With alternating current, high-voltage transmission lines can be made of smaller wires. This means that there will be less heat loss. Most generators deliver 13,200 volts. This voltage may be stepped up as high as 250,000 volts for long distance transmission and then eventually stepped down to the 120 or 240 volts that is used in homes and industry.

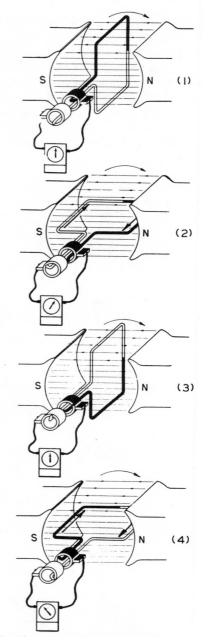

Fig. 19-18   Alternating Current Generator Rotation.

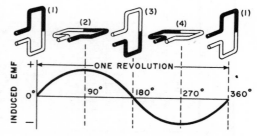

Fig. 19-19   Alternating Current Induced Electromotive Force Graph.

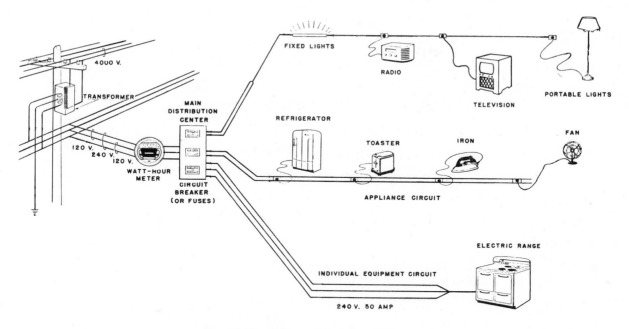

Fig. 19-20   A Typical Distribution System.

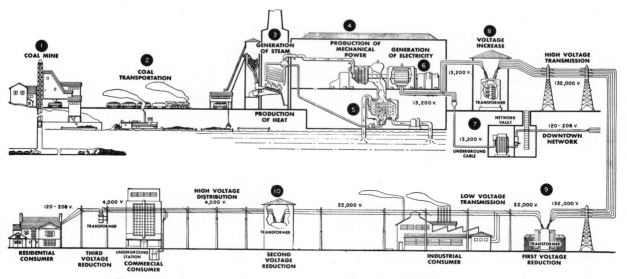

Fig. 19-21   Stages in Generating, Transmitting, and Distributing Electricity.

GENERAL STUDY QUESTIONS

1. Explain how a pencil or any other object is electrical in nature.

2. List the parts that make up an atom and tell the electrical characteristic of each.

3. What accounts for copper's ability to readily conduct electricity?

4. Why are elements whose atoms have almost complete outer shells poor conductors?

5. Do like charges of static electricity attract or repel each other?

6. Explain how a negatively charged rubber rod can charge an electroscope by contact.

7. Define an electrical current.

8. Define the following terms: amperage, voltage, resistance.

9. Using Ohm's Law, calculate the following:

    E - when I is 0.02 ampere and R is 4000 ohms

    I - when E is 110 volts and R is 220 ohms

    R - when E is 120 volts an I is 6 amperes

10. Draw a simple series circuit that contains three resistances.

11. Draw a simple parallel circuit that contains three resistances.

12. How can you account for the magnetic characteristics of iron and steel?

13. State the laws of repulsion and attraction for magnets.

14. How can the Left-Hand Rule be used to determine the direction of magnetic flux when the direction of current is known?

15. Describe how you could build a simple electromagnet.

16. Do electromagnets have any practical use? If so, list some common uses.

17. What is meant by primary and secondary cells?

18. What is electrochemical energy?

19. Explain the underlying principle of a generator.

20. Explain briefly how most generators are powered.

21. What is the basic difference in the construction of a d.c. and a.c. generator?

22. Why is alternating current used in preference to direct current for home and industrial uses?

23. Explain the function of a transformer.

CLASS DISCUSSION TOPICS

● Discuss the impact of electricity on our civilization.

● Discuss the electrical nature of matter.

● Discuss how electrons can move within a conductor.

● Discuss conductors and insulators.

● Discuss the meaning of static electricity.

● Discuss how an electroscope can be charged using a rubber rod; a glass rod.

● Discuss the relationship of voltage, amperes, and ohms - Ohm's Law.

● Discuss the theory of magnetism and electromagnetism.

● Discuss the electrochemical action that takes place in a dry cell or storage battery.

● Discuss the operation of d.c. and a.c. generators.

● Discuss the main advantage of alternating current.

● Discuss the generation, transmission, and distribution of electricity in your locality.

INTRODUCTION

Throughout this book many basic prime movers have been introduced and explained. These engines are capable of converting potential chemical energy (fuel) into kinetic mechanical energy — the energy of motion which can be easily observed at the end of the drive shaft. The kinetic mechanical energy of the engine can exert a force to move an object through a distance at a certain rate of speed which, of course, is POWER.

$$\text{Power} = \frac{\text{Force} \times \text{Distance}}{\text{Time}}$$

But we know that engines are not built to operate independently, merely as a tribute to the fine workmanship that is shown in their construction. The released chemical or heat energy that turns the shaft is not to be wasted — it is to be harnessed and controlled and put to man's use. Engines are used to power the machines that serve man so well.

DIRECT TRANSMISSION OF POWER

The power of the engine must be transmitted to these machines. In some cases, the "machine" can be fastened directly to the prime mover. This, to be sure, is desirable, simple, and few power losses occur along the way. When the drive shaft of the prime mover is directly connected to the driven shaft of the machine or device, an ideal situation exists as, for example, a blade fastened to the engine's crankshaft on a rotary lawnmower. Also, direct drive from the shaft can be seen in many machines that are driven by electric motors such as grinders and polishers. The simple, ideal situation for power transmission can be found, but in most cases the machine application calls for something more complex.

INDIRECT TRANSMISSION OF POWER

Direct transmission of power may not cover all of the operating requirements of the given machine or mechanism. Factors such as speed changes, disengaging the machine from the prime mover, torque requirements, changes in direction of rotation, transmitting energy over long distances, transmitting energy around corners. converting rotary motion to linear motion, and so forth, cannot be satisfied by direct, positive linkage with the prime mover. The power transmission system requirements of the automobile point up several of these factors. Appropriate speed changes and direction changes would not be practical if the crankshaft were directly and positively fastened to the wheels with no mechanism in between.

In addition to direct transmission of energy to machines, there are three primary methods of delivering the power of the prime mover to the machine that is to do the work.

1. Mechanical transmission which involves gears, pulleys, chain drive, levers, shafts, feed screws, etc., as the transmission media.

2. Fluid transmission involves the use of a liquid or a gas within a dynamic system as the transmission media.

3. Electrical transmission involves the use of conductive wire to deliver electric current. (Electrical transmission is covered in Unit 19.)

Each of the three methods has its advantages and disadvantages. Engineers select the transmission systems that fit the requirements and often they use the systems in combination with each other. For example: Coal is used to produce heat which converts water into steam which is directed against the blades of the steam turbine. The steam turbine drives a generator producing electrical energy which is transmitted throughout the countryside by the use of wires. The energy arrives at a factory, for instance, where it is used to drive various electric motors. The power produced by the electric motor may be transmitted to the machine through gears, belts, or other mechanical arrangements or the power of the electric motor may be transmitted to a fluid power system which, in turn, transmits the power to the machine. There is an infinite number of combinations. This example may awaken you to the field of power transmission. The remainder of this section will acquaint you with the physical principles and some of the techniques involved in transmitting power.

Unit 20

MECHANICAL TRANSMISSION OF POWER

Mechanical transmission principles enable a person to "do more" with the power that is available. The principles do not allow you to get any more out of the engine or other force-producing device; you cannot multiply the total power being produced but you can use the mechanical principles to adapt the power to a particular machine — you can become more flexible in what you do with the power.

Several mechanical principles have been applied for centuries — long before engines were a source of power. The principles work well whether the source of power is an engine, a draft animal, a human, or a natural force. The principles discussed in this unit are often referred to as "simple machines." These principles are restated by engineers in thousands of variations as seen in transmission systems and the machines themselves. In fact, it is sometimes difficult to draw a hard and fast line between the transmission system and the machine itself; the two may often blend into an integral unit. Countless applications of the simple machine principles are seen but basically the principles can accomplish these things:

1. Change the size of the applied force.

2. Change the direction of the applied force.

3. Change the speed that results from the applied force.

No doubt you will quickly recognize many of the simple machines we shall discuss. We all come in contact with their principles every day: Lever, Wheel and Axle, Inclined Plane, Wedge, and Screw.

LEVER

Levers enable a person to lift or move or hold far more than he could unaided. They are commonly seen: a teetertotter enables a small child to lift his mother off the ground, a man uses a lever to move a boulder, a girl uses a pair of pliers to hold a bolt. Any long shaft or board will do for a lever as long as you have a pivot point or fulcrum to work against. In gaining a mechanical advantage with a lever, you use less force but you must apply it over a greater distance to overcome and move the opposing object which resists movement.

Mechanical advantage is the number of times that force is multiplied. What you gain in applied force you lose in having to apply the force over a greater distance. The principle of the lever will not allow you to have both larger force and smaller distances moved; one is sacrificed for the other. With the lever shown in Fig. 20-1, an effort or force of 25 pounds can lift 100 pounds; hence, a mechanical advantage of four.

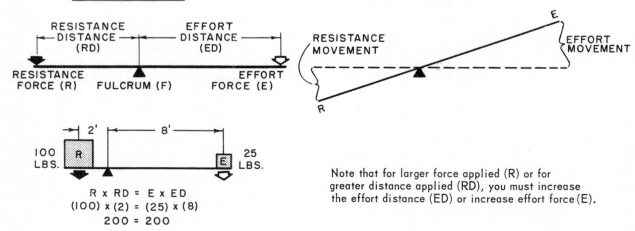

$$R \times RD = E \times ED$$
$$(100) \times (2) = (25) \times (8)$$
$$200 = 200$$

Note that for larger force applied (R) or for greater distance applied (RD), you must increase the effort distance (ED) or increase effort force (E).

Fig. 20-1 The Moments of Force Around the Fulcrum Balance Each Other.

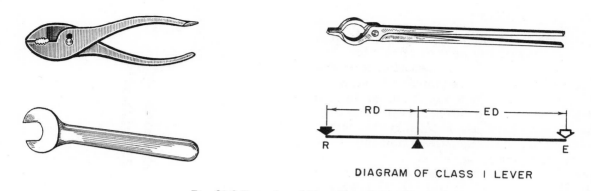

Fig. 20-2 Examples of First Class Levers.

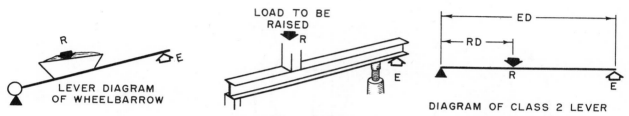

Fig. 20-3 Examples of Second Class Levers.

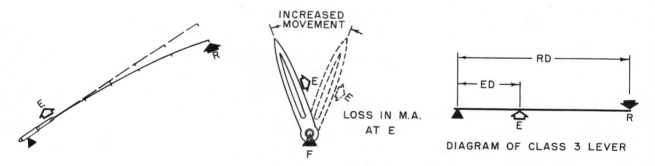

Fig. 20-4 Examples of Third Class Levers.

Levers fall into three categories — first class levers, second class levers, and third class levers.

First class levers place the fulcrum between the resistance and the effort force. These levers multiply force with a sacrifice in the distance the object is moved. The direction of movement is reversed.

Second class levers place the fulcrum at the end of the lever with the resisting weight at a point along the lever. These levers also multiply force with a sacrifice in the distance the

object is moved. The direction of object movement is the same as that of the applied force.

Third class levers place the fulcrum at the end of the lever with the resisting weight at the opposite end of the lever. The applied force or effort is at a point along the lever. These levers increase effort but multiply distance moved by the object and therefore have a negative mechanical advantage. A fishing rod illustrates this principle and sometimes the fisherman is mystified at how a small fish can create such a large pull on his rod. The fish has the mechanical advantage.

WHEEL AND AXLE

Essentially the wheel and axle consists of two wheels firmly held together along the same axis. The axle may be short or it may be a very long shaft. Mechanical advantage is gained when the large wheel is moved through a distance. The axle moves also; its motion is less since its diameter is less, but the force, twist, or torque on the axle is multiplied. A boat trailer winch is an example of this principle. The mechanical advantage of the winch will allow a man of normal size to properly load and secure a heavy boat onto a trailer.

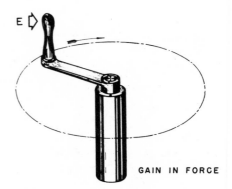

Fig. 20-5 Gaining Mechanical Advantage of Force with the Wheel and Axle.

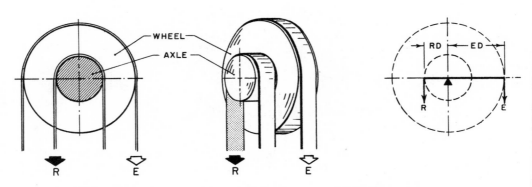

Fig. 20-6 Wheel and Axle Moments of Force. How Would a Mechanical Advantage of Speed Be Gained?

Study Fig. 20-6. Note that the mechanical advantage possible with the wheel and axle is essentially that of a Class 1 lever. The fulcrum is at the center of the wheel and axle with the resistance distance (RD) being measured from the fulcrum to the periphery of the axle and the effort distance (ED) being measured from the fulcrum to the periphery of the wheel.

In the perfect wheel and axle these moments of force are equal: $E \times ED = R \times RD$.

Fig. 20-7 shows an effort force of ten pounds applied to a five-inch radius wheel. The resistance force acting on the axle of $1''$ radius which the effort force overcomes is found by the moment formula:

$$E \times ED = R \times RD$$

$$10 \text{ pounds} \times 5'' = R \times 1''$$

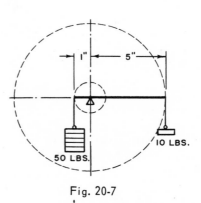

Fig. 20-7

$$\frac{50 \text{ pound-inches}}{1 \text{ inch}} = R$$

$$50 \text{ pounds} = R$$

It is apparent that with a wheel and axle combination a smaller force may be used to produce or control a larger force.

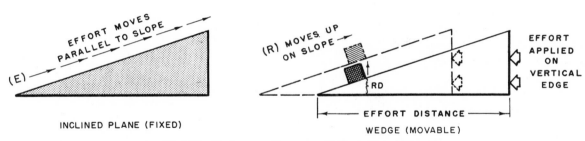

Fig. 20-8 The Wedge is a Variation of the Inclined Plane.

INCLINED PLANE

The inclined plane also allows force to be applied over a longer distance in order to handle heavy objects. This principle as well as others was used by the Egyptians in the construction of the pyramids. Stone blocks weighing several tons were rolled up a gently inclined plane and maneuvered into place. Upon completion of the structure, the earthen inclined planes were removed.

We are familiar with ramps. Let us see how they allow us to apply a lesser amount of force over a longer distance to ease the job at hand.

Example: A boy lifts his 100-pound body onto a three-foot loading dock with a single jump.

Work = Force × Distance

Work = 100 lbs. × 3 ft. = 300 ft.-lbs.

$$\text{Force} = \frac{\text{Work}}{\text{Distance}} = \frac{300 \text{ ft.-lbs.}}{3 \text{ ft.}} = 100 \text{ lbs.}$$

By walking up a ten-foot long ramp he can apply less than the 100-pound force required in a single jump but he must apply it through a greater distance.

$$\text{Force} = \frac{\text{Work}}{\text{Distance}} = \frac{300 \text{ ft.-lbs.}}{10 \text{ ft.}} = 30 \text{ lbs.}$$

Force = 30 lbs. when boy walks up ramp

In this example, the boy actually has a mechanical advantage of force when he uses the inclined plane. His mechanical advantage of force is 3 and 1/3.

$$MA_f = \frac{R \text{ (Resistance Force)}}{E \text{ (Effort Force)}}$$

$$MA_f = \frac{100 \text{ lbs.}}{30} = 3 \; 1/3$$

The mechanical advantage of distance can also be calculated.

$$MA_d = \frac{RD \text{ (Resistance Distance)}}{ED \text{ (Effort Distance)}}$$

$$MA_d = \frac{3 \text{ ft.}}{10 \text{ ft.}} = .3$$

The mechanical advantage of speed is calculated with the same formula:

$$MA_s = \frac{RD}{ED}$$

Note that these mechanical advantage formulas apply to all simple machines.

DETAIL OF
INSIDE CUTTER

Fig. 20-9 Applications of the Wedge as a Separating Device.

209

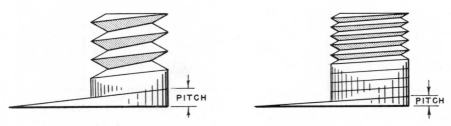

Fig. 20-10 The Screw, an Inclined Plane in Spiral Form.

WEDGE

The wedge is a variation of the inclined plane. It is driven under or into an object and applies its force through a longer distance. Many cutting tools are actually made up of a series of wedges which we call teeth. Under magnification these wedge teeth "get under" and break away the material they are cutting. Axes and chisels are also wedges. In addition, wedges make excellent holding devices, the simple doorstop being but one example.

SCREW

The screw is also an inclined plane. However, it is one that has been put into a round form and appears as a continuous spiral. The most common application is in fasteners such as nuts and bolts and wood screws. However, screw principles are used in jacks, worm gears, some pumps, and to convert rotary motion to linear motion. The mechanical advantage that can be applied by a screw can be tremendous since the pitch of the screw is usually quite small (RD) and the distance the effort is applied through (ED) is usually quite large by comparison. A jack screw illustrates this very well.

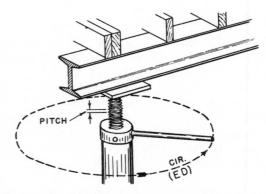

Fig. 20-11 The Jack Screw Applies a Tremendous Mechanical Advantage of Force.

PULLEYS

Pulleys in various arrangements can create a mechanical advantage which can aid in the movement of heavy objects. In gaining the mechanical advantage, distance moved is increased and sacrificed in the interest of being able to apply less force. The number of ropes supporting the object determines the mechanical advantage. If four ropes support the weight, then the rope must be pulled 4 feet in order to move the object 1 foot, but the force (disregarding friction) required would be one-fourth the weight of the object.

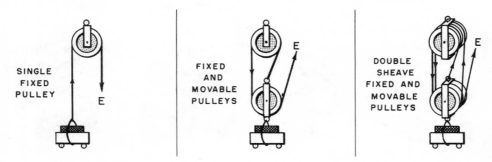

Fig. 20-12 Pulleys Increase in Mechanical Advantage as the Number of Turns of Rope (ED) Increases.

GEARS

The gear is very common and has not been discussed up to this point, but gears use the same principles as do simple machines. The gear is simply a wheel with special notches or teeth cut into its perimeter to provide positive contact with another gear. In a sense, two meshing gear teeth are like two levers working against each other.

V-belts and pulleys or chains and sprockets have the same mechanical function as do gears and they are very useful if power must be transmitted between shafts that are separated by a distance. If construction permits close spacing of the shafts, then gear teeth can mesh against each other, a simpler arrangement.

Gears of different sizes change speed in proportion to the number of teeth on each gear. A gear with 24 teeth will drive a gear with 12 teeth at double its speed. A gear with 24 teeth will drive a gear with 48 teeth at half its speed.

Fig. 20-13 A Chain-Driven Gear Train.

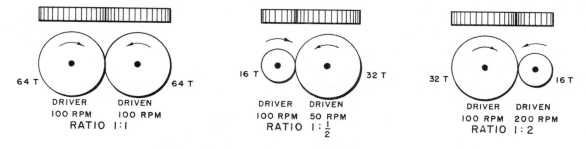

Fig. 20-14 Gears May Be Used to Change Speeds.

$$MA_f = \frac{\text{Teeth in Driven Gear}}{\text{Teeth in Driving Gear}} \qquad MA_s = \frac{\text{Teeth of Driving Gear}}{\text{Teeth in Driven Gear}}$$

When speed changes, the torque or twisting force also changes. Gearing up to increase speed will reduce the torque produced; gearing down to decrease speed will increase torque. The gearing in an automobile is an excellent example of this. At low speed you gear down to gain the torque needed to get the auto moving. At highway speed, the auto is moving nicely and you gear up — less torque is needed, but more speed is required. The power transmission principles of the automobile are discussed in more detail in Unit 13 — The Automobile Engine.

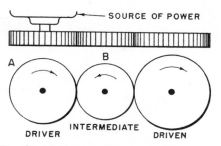

Fig. 20-15 Changing Direction of Rotation.

Gears also change the direction of rotation as the power is transmitted from one to another; from clockwise rotation to counterclockwise rotation and so forth.

Gears are arranged in a gear box in the proper sizes and combinations to satisfy the needs of the machine that is to be powered. Sometimes the positions and gear combinations are fixed but often provision is made to shift the gears around in various combinations as is done in an automobile transmission. Gears in no way change the total power that is developed by the prime mover — gears just change the speed or torque.

If the positions of the gears are to be changed while the prime mover is operating, the gears must be momentarily disengaged from the drive shaft while the gears are shifted in position. A clutch serves this purpose.

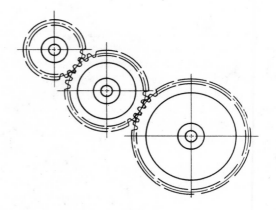

Fig. 20-16 Representation of Simple Gear Train.

UNIVERSAL JOINTS

Universal joints add flexibility to the mechanical transmission of power in that they allow transmission between shafts that do not share the same axis. They do have practical limits — they cannot go around corners or in circles but in certain applications they are very useful. The human wrist is a version of the universal joint and can serve to illustrate its action.

In addition to universal joints, flexible couplings can be used to transmit power at an angle to the prime mover shaft. Flexible shafts can also accomplish this.

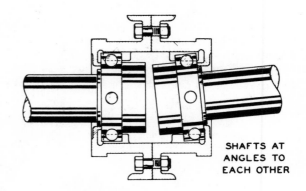

SHAFTS AT ANGLES TO EACH OTHER

Fig. 20-17 Flexible Coupling.

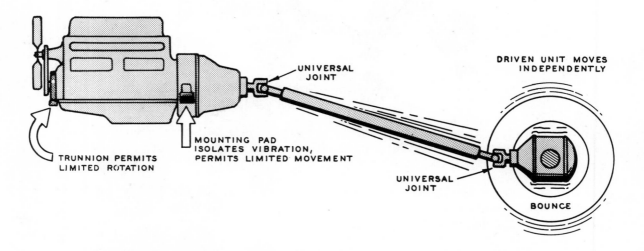

UNIVERSAL JOINT

DRIVEN UNIT MOVES INDEPENDENTLY

TRUNNION PERMITS LIMITED ROTATION

MOUNTING PAD ISOLATES VIBRATION, PERMITS LIMITED MOVEMENT

UNIVERSAL JOINT

BOUNCE

Fig. 20-18 Universal Joints Add Flexibility to Mechanical Power Transmission Systems.

SUMMARY

Mechanical transmission of power can be seen and analyzed in everyday machines and mechanisms. Understanding some of the basic principles of machines and power transmission will add to a person's knowledge and also will take some of the mystery out of devices that at first view appear to be too complicated to be understood. Mechanical transmission is the world of gears, wheels, levers, belts, pulleys, chains, shafts, and screws. They are positive linkages between the prime mover and the mechanism. Through years of improvement and research they have become very efficient and reliable in their service.

GENERAL STUDY QUESTIONS

1. Can direct or fixed transmission of power between the prime mover and the machine satisfy all machine applications? Explain.

2. What are the three basic power transmission systems?

3. List practical applications of first, second, and third class levers.

4. Calculate the force necessary to move a 200-pound weight using a first class lever. The weight is 3 feet from the fulcrum. The lever's total length is 13 feet.

5. Calculate the mechanical advantage of force found in carrying a 50-pound box up a 10-foot ramp that is 2 feet from ground level.

6. Compare a "conventional" wedge to the wedge formed by a cutting tooth.

7. What could be said about the mechanical advantage that is created by the use of the screw as a simple machine?

8. What determines the mechanical advantage of a pulley?

9. Why would V-belts or chain drive be used instead of gear drive?

10. Explain the relationship that exists between speed and torque.

11. What is the mechanical advantage of speed between a driven gear with 12 teeth and a driving gear with 96 teeth?

12. List several machine applications that might require a universal joint, a flexible joint, or a flexible shaft.

CLASS DISCUSSION TOPICS

- Discuss and cite examples of how different power transmission methods can be combined in a single system.

- Discuss why a transmission system must be flexible in most applications.

- Cite and discuss examples of power transmission systems that are an integral part of the machine.

- Discuss the relationship between the wheel and axle and a Class 1 lever.

- Discuss the terms: mechanical advantage of force, mechanical advantage of distance, and mechanical advantage of speed.

- Discuss how a feed screw can convert rotary motion into linear motion.

Unit 21

FLUID TRANSMISSION OF POWER

The term "Fluid Power" has a rather modern or contemporary sound to it but actually fluid power has deep roots in history. By definition, fluid power involves the use of a fluid, either liquid or gas, to operate a mechanism. The fluid might be air-directed as wind captured by the sail of a ship, or the fluid might be water flowing in a river, its energy tapped by a water wheel. Further, we know that in the civilizations of the Romans, Cretans, Persians, Indians, and Chinese, water has been stored in reservoirs, channeled in aqueducts and pipes, and used for domestic purposes and for irrigation. Men did not progress beyond these applications of fluid power for many centuries and certainly these are not the techniques of application that concern today's technology.

However, before bringing the fluid power story into the focus of modern times and modern applications, it would be wise to clearly understand the two basic terms that are associated with fluid power: (1) Hydraulics and (2) Pneumatics.

Hydraulics is that branch of fluid power that deals with the use of liquids (usually oil) to transmit power from one place to another. The hydraulic system receives and controls the power delivered by the prime mover, generates a hydraulic force, transmits the force, and converts the force to power which is delivered to the work load.

Pneumatics is that branch of fluid power that deals with the use of gases (usually air or nitrogen) to transmit power from one place to another. The pneumatic system receives and controls the power delivered by the prime mover, generates a pneumatic force, transmits the force, and converts the force to power which is delivered to the work load.

Hydraulics and pneumatics share many common characteristics:

1. Both use a fluid media, a liquid will take the shape of the vessel in which it is contained and a gas, in addition, will completely fill the vessel in which it is contained.

2. Both liquids and gases can flow through a pipe or a hose.

3. The rate of flow of both liquids and gases can be easily controlled with valves.

4. Both liquids and gases can be caused to flow by various types of pumps.

There is one basic difference between liquids and gases which makes the separation of the two areas into different study units easier to follow. Liquids are virtually incompressible while gases can be compressed. A hammer blow on the piston of a liquid-filled cylinder is like hitting the end of a steel shaft. A hammer blow on the piston of a gas-filled cylinder is like hitting a spring-loaded cylinder — it gives and then bounces back. Many similarities exist between hydraulic systems and pneumatic systems but for the sake of clarity the two will be discussed separately.

HYDRAULICS

The key principle for modern hydraulics was formulated by Blaise Pascal in the 17th century. Pascal theorized that: "If a vessel full of water, and closed on all sides, has two openings, one a hundred times as large as the other, and if each is supplied with a piston that 'fits exactly,' then a man pushing the small piston will exert a force that will equiliberate that of 100 men pushing on the large piston and will overcome that of 99 men."

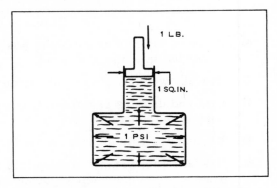

Fig. 21-1 Pressure is Transmitted Equally in All Directions.

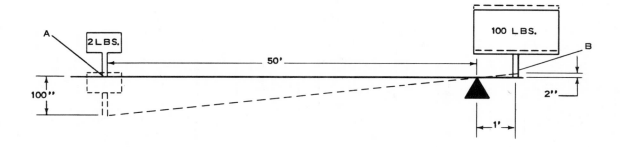

Fig. 21-2   The Mechanical Advantage of a Lever Also Applies to Hydraulics.

Sometimes this law is simply stated: "The pressure at any point in a static liquid is the same in every direction and exerts equal force on equal areas," Fig. 21-1.

Pascal's law is really an application of mechanical advantage to a liquid. Just as a lever can multiply force at the sacrifice of distance, so can hydraulics.

In the illustration of the lever and the weight, Fig. 21-2, a force of 2 pounds moves through 100 inches to produce 200 inch-pounds of work. (Work = Force × Distance.) This work is capable of producing equal work on the other end of the lever and can move 100 pounds through a distance of 2 inches, or 200 inch-pounds.

The hydraulic creation of mechanical advantage is similar to that of the lever, Fig. 21-3. Two pounds of force pushing the piston down the cylinder 50 inches is equal to 100 inch-pounds of work. The two pounds of force is transmitted undiminished in all directions and pushes against the 50 square inches of the large piston with a total force of 100 pounds. And, since Work = Force × Distance,

$$\text{Distance} = \frac{\text{Work}}{\text{Force}}$$

$$= \frac{100 \text{ inch-pounds}}{100 \text{ pounds}}$$

$$= 1 \text{ inch}$$

The large piston moves upward 1 inch.

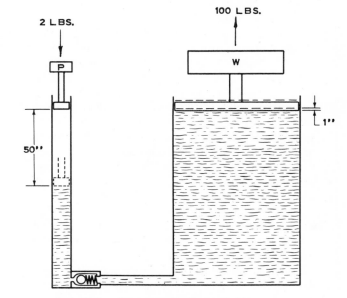

Fig. 21-3   Creating a Mechanical Advantage in Hydraulics.

Pascal's law deals with static liquids, that is, liquids that are not in motion. Static pressure is one thing, the dynamics of liquids or liquids in motion is quite another. As was indicated earlier, friction was not taken into account. But friction is a factor in fluid transmission of power just as it is in the mechanical transmission of power. Friction in dynamic fluids is always present but it can be minimized if: (1) the pipes or hoses are large enough to prevent excessive velocity and the resulting turbulence, (2) the lengths of the pipes or hoses are not excessive, and (3) there are not an excessive number of bends in the pipe and the corners are not too sharp.

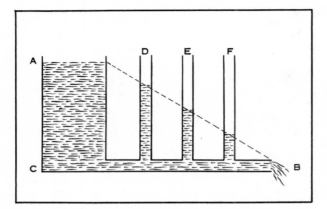

Fig. 21-4   The Drag of Friction Also Applies
to Liquids.

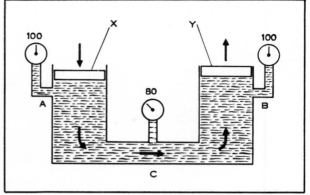

Fig. 21-5   Static Pressure Decreases as
Velocity Increases.

The principle of how friction affects liquid flow can be seen in Fig. 21-4. The main body of the liquid between A-C has created what is called a static head pressure of . . . let's say 20 p.s.i. (pounds per square inch). When point B is opened up, the water is free to flow and dissipate the entire pressure. Notice that the water does not rise to the level of A in pipes D, E, and F; it is noticeably less the further the pipes are from the source. The drag of friction prevents the water from rising in these pipes; friction has robbed us of some of our static head pressure.

Turbulence is another factor in fluid dynamics. A smooth laminar (in layers) flow of liquid is the most efficient. The layers of moving fluid are parallel to each other, layers near the center of the pipe are moving more rapidly than those near the surface of the pipe where friction creates some drag. However, if the velocity becomes too great due to a pipe that is too small, the laminar flow is disturbed and then efficiency-destroying turbulence sets in. Laminar flow might be compared to the smooth flow of traffic on a super highway under moderate traffic load. Rush hour traffic slows, it is more turbulent, and therefore slower and less efficient.

Bernoulli's principle also concerns itself with the dynamic behavior of fluids. This principle states that "the static pressure of a moving liquid varies inversely as its velocity." The static pressure on the surface of the pipe or hose decreases as the velocity of the fluid increases. This principle is applied in carburetors when the fluid flowing is air, rushing

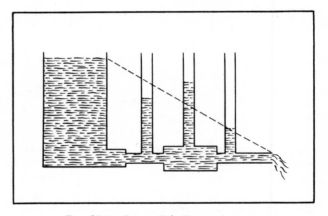

Fig. 21-6   Bernoulli's Principle in Action.

through the venturi section. Low pressures are created in the venturi section. Figs. 21-5 and 21-6 show how static pressure is reduced in constricted areas where the velocity is increased.

These are some of the principles that provide guidelines for engineers to follow as they design various hydraulic units that make up hydraulic systems of power transmission. Hydraulic systems are commonplace today but in terms of long span history, hydraulics is a young industry. Modern hydraulics is not water flowing in aqueducts — it is liquid under high pressure in high pressure pipes and hoses and in close tolerance machines and leak-proof pistons and cylinders. The development of reliable fluid power transmission systems had to wait for improved materials and manufacturing techniques before it could become a reality.

HYDRAULIC COMPONENTS

The layman may even be unaware of the many hydraulic systems with which he comes in contact with each day. He may be aware of some uses such as the automobile brake system, power steering, a garage hoist, or a dump truck. But he may not realize the wide application of hydraulics to the <u>Manufacturing Industries</u> such as machine tools, food and chemical processors, die casting, plastics, presses, glass machinery, textile mills, paper products, ordnance, deck machinery, or applications in the <u>Aero Space Industries</u> such as commercial aircraft, military aircraft, rockets, satellites, helicopters, or applications for <u>Mobile Equipment</u> such as lift trucks, earth movers, farm machinery, log skidders, trucks and buses, utility vehicles, mining, and railroad equipment. No doubt this list has pointed to applications that you may well be familiar with but had actually not given much thought.

But exactly what is a hydraulic system? What parts are needed? Just how is it able to transmit power from a prime mover or electric motor to the machine or mechanism? The components of a hydraulic system that answer these questions include:

1. <u>Reservoir,</u> or storage tank for liquid.

2. <u>Tubing, pipes</u> or <u>hose</u> to transmit the liquid through the hydraulic circuit.

3. <u>Pump,</u> driven by the prime mover in order to move the hydraulic fluid under pressure.

4. <u>Valves,</u> to change direction of flow, pressure of flow, or amount of flow.

5. <u>Motor</u> or <u>cylinder,</u> which is acted on by the hydraulic fluid, converting hydraulic power back into mechanical power for use by the mechanism.

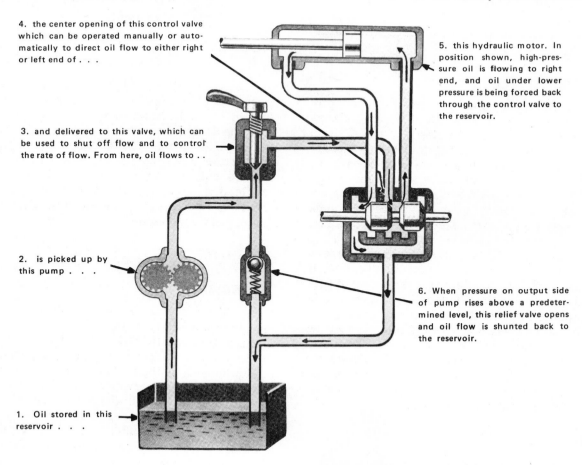

4. the center opening of this control valve which can be operated manually or automatically to direct oil flow to either right or left end of . . .

5. this hydraulic motor. In position shown, high-pressure oil is flowing to right end, and oil under lower pressure is being forced back through the control valve to the reservoir.

3. and delivered to this valve, which can be used to shut off flow and to control the rate of flow. From here, oil flows to . .

2. is picked up by this pump . . .

6. When pressure on output side of pump rises above a predetermined level, this relief valve opens and oil flow is shunted back to the reservoir.

1. Oil stored in this reservoir . . .

Fig. 21-7 A Typical Hydraulic System.

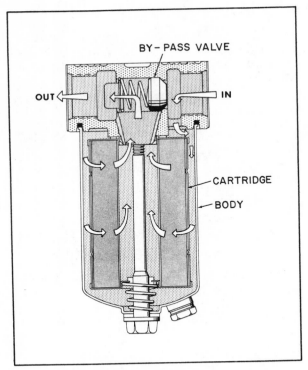

Fig. 21-8   Full Flow Oil Filter.

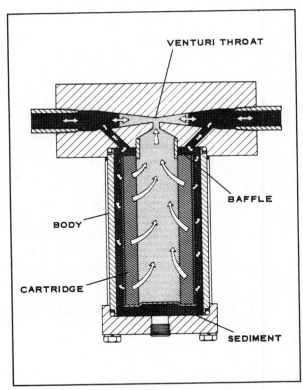

Fig. 21-9   Proportional Oil Filter.

HYDRAULIC FLUID

The fluid most often used in hydraulic systems is oil. Hydraulic oil is formulated especially to satisfy the needs of the system. Its viscosity is such that it flows freely through the system at the system's designed operating temperature. It must not be so thin that it can leak from the system and reduce efficiency but on the other hand it must be sufficiently thick to lubricate the various components properly. Over a period of time oil may begin to oxidize or combine with air just as other substances do. Oxidization must be controlled because it causes the development of resins, varnishes and lacquers, and sludges. High operating temperatures increase the rate of oxidization in oil. The contaminants caused by oxidization are acid in nature and corrosive. Corrosive effects on the precision hydraulic components can be minimized by using equipment that works at moderate pressure and temperature and by careful preventative maintenance.

FILTERS

Oil filters are a part of every system and are located within the reservoir. Normal amounts of oxidization products or foreign particles can be handled by a filter. The filter may be designed as a full flow filter, Fig. 21-8, in which all of the oil that enters the pump is filtered or it may be a proportional filter which filters only a portion of the oil during each cycle, Fig. 21-9. Many filters are designed to remove particles that are much smaller than the human eye can detect.

RESERVOIR

The reservoir itself is a steel tank that is large enough to hold a sufficient reserve supply of hydraulic oil. Most reservoirs are vented to the atmosphere to prevent a vacuum from forming as the oil level varies. The vent is fitted with an air breather or filter to prevent foreign materials from entering the system. As the oil returns to the reservoir, it enters with a good deal of force and turbulence; baffle plates slow down the oil and allow time for any air bubbles to rise to the top before the oil is recirculated.

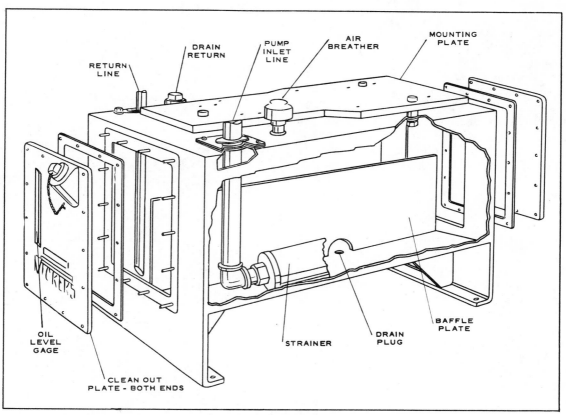

Fig. 21-10   A Typical Reservoir.

PUMPS

The hydraulic pump is driven by the prime mover or electric motor. It converts mechanical energy into hydraulic energy. It picks up the hydraulic fluid and forces it along its way. Often people want to think of these pumps as building up an oil pressure, but since oil is virtually incompressible, it is better to think of the pump as delivering a driving rod of solid oil through the piping of the system. The pump generates flow and the resulting force. Pressure or resistance to the flow occurs further down the line as the machine places a load against the hydraulic force.

Pumps are rated or sized by the volume of hydraulic fluid that they can deliver within a given time — gallons per minute. This output varies with the speed of the pump so the pump speed must be established as a factor in a pump rating. Some pumps are rated by their displacement or the amount of oil they will move during one cycle. Displacement is measured in cubic inches.

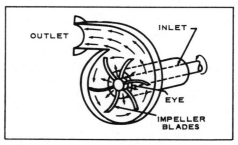

Fig. 21-11   Operating Principle of a Centrifugal Pump which is a Nonpositive Displacement Pump.

Pumps operate on either of two basic principles: (1) Nonpositive displacement and (2) Positive displacement.

Nonpositive Displacement Pumps

The nonpositive displacement pump is characterized by the centrifugal pump. The speed of the pump and the resistance to flow determine how much fluid is delivered. The principle is simply that of centrifugal force throwing the fluid from the eye of the pump where it enters, to the outside periphery of the spinning impeller and through the pump outlet. If the pump discharge were completely shut, the pump would churn through the liquid, producing heat. One hundred percent slippage is possible with the centrifugal pump. These pumps are well adapted for applications that require a large volume of flow at relatively low pressures.

Positive Displacement Pumps

Positive displacement pumps deliver a fixed volume of fluid at any given speed. If resistance to flow is increased, the force of the liquid and its velocity is increased. If the resistance became absolute or the flow is cut off, then pressure would build up to the stall or breaking point if it were not for a relief valve which will open at a predetermined pressure setting. There is little if any slippage in a positive displacement pump — hence the term "positive."

The pump speed is usually not varied to meet differing operating requirements. If the application calls for flexibility in the volume of the fluid flow, then variable displacement pumps are used. In these pumps the physical relationship between the pump components can be changed to provide for different volume requirements. Flow control valves, which are discussed later, can also change the volume of fluid flow.

Rotary pumps are of the positive displacement type and include gear pumps, lobe pumps, vane pumps, and piston pumps. In any of these pumps the hydraulic fluid is allowed to enter the pump chamber, trapped by the closing pump chamber and then moved to the discharge side of the pump where it is forced out. The key parts may be gears rotating against each other, lobes rotating against each other, movable vanes that work against the walls of the pump housing, radial pistons in a rotating off-center cylinder block, or axial pistons in an offset or tilted cylinder block. Several simplified drawings of these pumps are shown, Fig. 21-12, A through E.

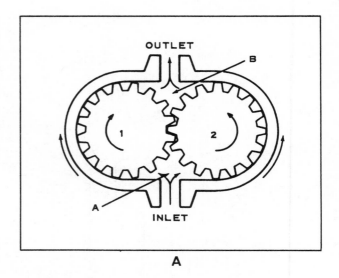

A

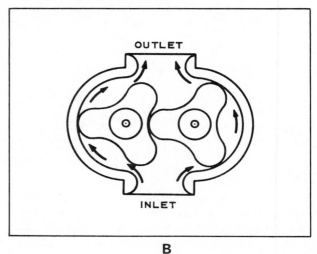

B

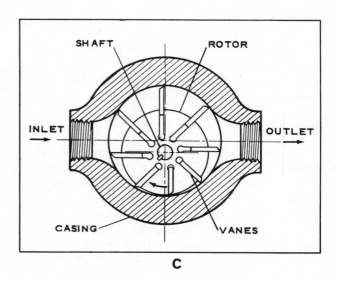

C

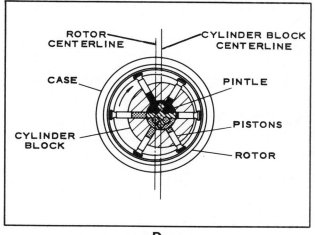
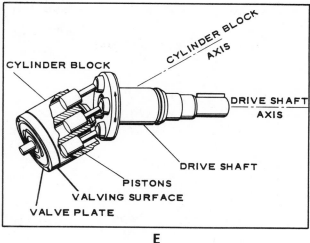

D   E

Fig. 21-12   Positive Displacement Hydraulic Pumps: (A) Gear Pump, (B) Lobe Pump,
(C) Vane Pump, (D) Radial Piston Pump, and (E) Axial Piston Pump.

TUBING

Fluid from the pump is transmitted through pipes, tubing, or flexible hose. These transmitting lines must be strong enough to contain the hydraulic fluid at its working pressure and accept higher surges in pressure should they occur. Often hydraulic lines must contain fantastically high pressures — up to 5000 lbs. per sq. in. are used in some systems.

VALVES

Valves are located in the hydraulic circuit to control the rate of fluid flow, the pressure of the system or the direction of the fluid flow — rate, pressure, or direction valves.

Flow Control Valves

Rate valves, commonly known as flow control valves, operate on valve principles that are rather well known and commonly used in handling many different fluids. Rate valves include gate, plug, globe, spool, and needle valves, Fig. 21-13. Gate, plug, and spool valves are normally ON-OFF valves; they are not suited for use as a throttle since in a partially opened position they cause excessive turbulence. The globe valve is also normally ON or OFF but it can be used as a throttle since the liquid flows on all sides of the globe and it is more balanced in operation. Needle valves are well suited to vary the amount of flow through small openings.

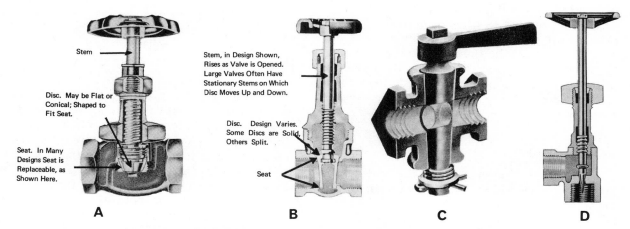

A   B   C   D

Fig. 21-13   Typical Flow Control Valves: (A) Globe Valve, (B) Gate Valve,
(C) Plug Valve, and (D) Needle Valve.

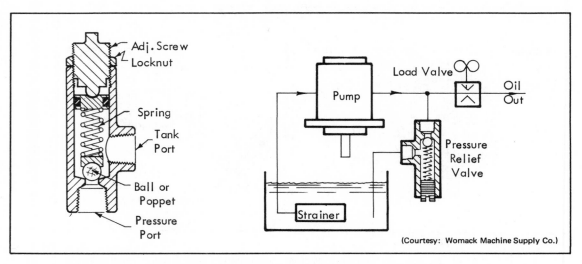

Fig. 21-14   Pressure Relief Valve and Its Application.

Pressure-compensated flow control valves accomplish the same basic task as a governor does on an engine. These valves provide a constant pressure and hence a constant speed at the working machine, even though the load on the machine may vary. A spring-loaded piston moves to increase the valve opening when the pressure begins to drop at the discharge side of the valve. The pressure balances the spring to provide a uniform rate of fluid flow.

Relief Valves

Pressure relief valves, Fig. 21-14, are set to open at a predetermined pressure; they are a safety feature which prevents excessive pressure from building up. They are spring-loaded; the valve is closed when the spring pressure is equal to or greater than the pressure of the hydraulic fluid. At fluid pressures higher than the spring pressure, the valve is pushed open and a portion of the fluid is bled off, thereby reducing the force or pressure in the system.

Reducing Valves

Another type of valve is the pressure-reducing valve which allows a branch of the main hydraulic circuit to operate at a lower pressure than the main circuit pressure. These are spring-loaded valves which bleed off pressure to balance the valve at a preset pressure.

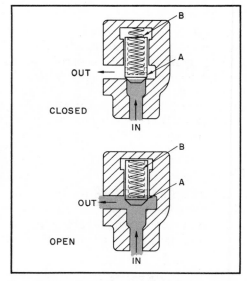

Fig. 21-15   Check Valve Allows Flow in One Direction Only.

Directional Valves

The check valve is perhaps the most common directional valve. This valve permits flow in one direction only; as soon as pressure in the opposite direction is applied, the valve mechanism is forced tightly against its seat. In Fig. 21-15, the check valve is a spring (B) loaded piston (A). If the force of fluid flowing through the valve is stopped, the spring pushes the piston against the valve seat preventing any fluid flow through the valve in the opposite direction.

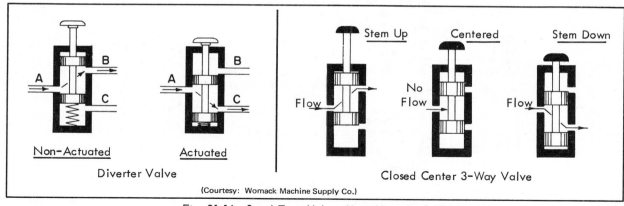

(Courtesy: Womack Machine Supply Co.)

Fig. 21-16   Spool Type Valves Have Many Applications.

Spool valves are manufactured in a wide variety and they are capable of controlling the direction of flow, Fig. 21-16 and Fig. 21-17. The liquid can flow around the spool shaft when and where a path is provided.

Shifting the spool shaft back and forth can open and close the various paths. These valves, like many others, can be located in remote positions and can be actuated by solenoids.

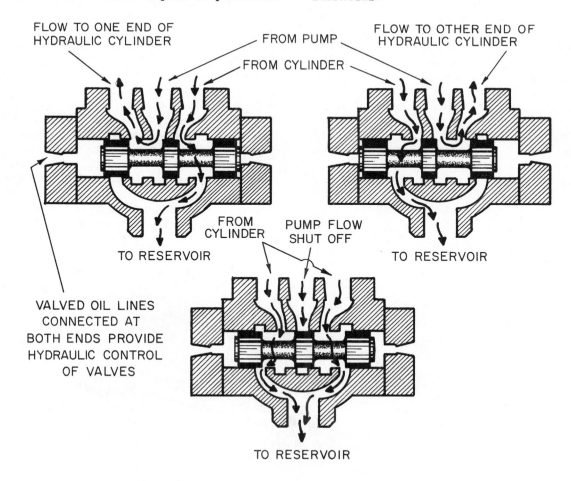

Fig. 21-17   Spool Valves are Directional Valves that Provide Quick, Positive Action.

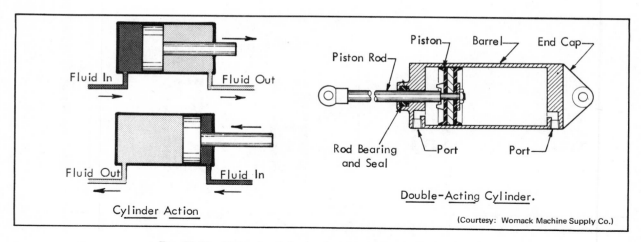

Fig. 21-18 Hydraulic Cylinder Action and Basic Construction.

MOTORS

The hydraulic fluid now arrives at the motor where hydraulic energy is reconverted into mechanical energy. If linear motion is the desired output of the system, a hydraulic cylinder is used as the motor. If rotary motion is the desired output, a rotary motor is used. Regardless of which motor is used, its output is converted into power that moves against the work load at the machine. Rotary motors perform just the opposite function as do rotary pumps, just as an electric motor performs just the opposite function of the electrical generator. In many cases, pumps can be used as motors with little or no modification. Motors are selected on the basis of the speed and torque that are required.

HYDRAULIC CYLINDER ACTION

In a hydraulic cylinder, Fig. 21-18, the fluid applies a force in all directions inside the cylinder but only the piston is free to respond to the force which it does by sliding down the cylinder — this linear motion of the piston producing work at the machine. When the cylinder has completed its stroke, the fluid direction is changed and hydraulic force on the opposite end of the piston moves the piston back to its original position. The area of the piston face has a good deal to do with the speed of the piston travel; here again we have our choice of speed and long distance or larger force. As the size of the piston increases, the speed decreases and the force is applied through a shorter distance.

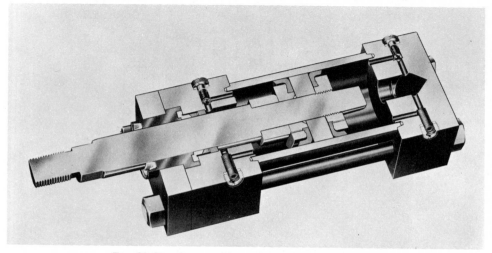

Fig. 21-19 Cutaway View of a Hydraulic Cylinder.

If the efficiency of a cylinder is to be high, it must be well sealed at both the rod and the piston. The development of superior packing and sealing materials was a problem in the hydraulics industry for many years but modern materials prevent leakage within the cylinder. There are a surprising number of applications for cylinders; several are illustrated, Fig. 21-20.

SUMMARY (Hydraulics)

This, then, is the complete hydraulic system: it consists of the (1) Reservoir, (2) Pump, (3) Transmission Lines, (4) Pressure, Flow, and Directional Valves, and (5) Motor. Fig. 21-21 shows a typical hydraulic system and it should be meaningful to you at this point in your studies.

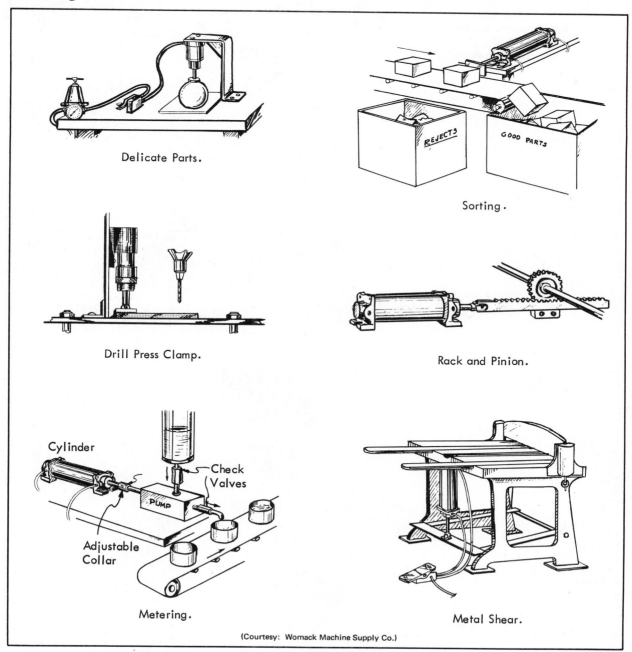

Delicate Parts.

Sorting.

Drill Press Clamp.

Rack and Pinion.

Cylinder

Check Valves

PUMP

Adjustable Collar

Metering.

Metal Shear.

(Courtesy: Womack Machine Supply Co.)

Fig. 21-20   A Few of the Many Applications for Hydraulic Cylinders.

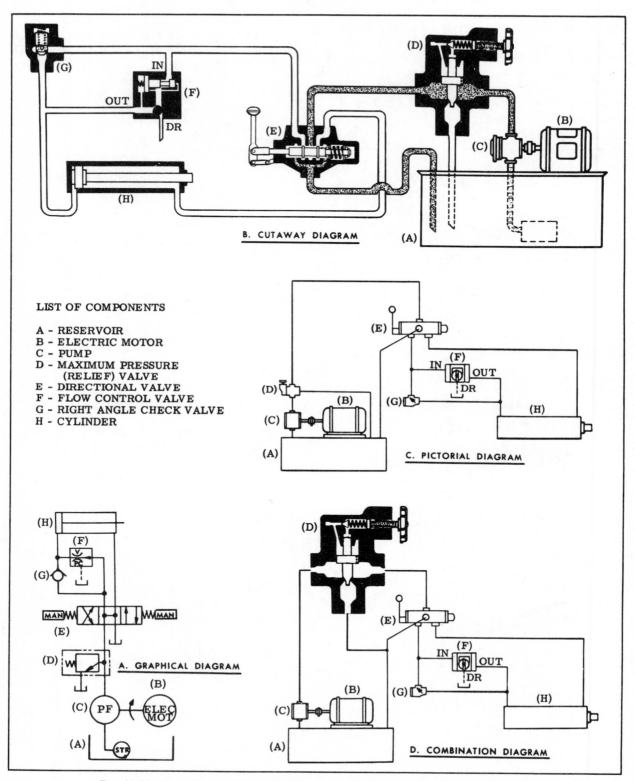

LIST OF COMPONENTS

A - RESERVOIR
B - ELECTRIC MOTOR
C - PUMP
D - MAXIMUM PRESSURE
      (RELIEF) VALVE
E - DIRECTIONAL VALVE
F - FLOW CONTROL VALVE
G - RIGHT ANGLE CHECK VALVE
H - CYLINDER

Fig. 21-21   Four Types of Hydraulic Drawings: (A) Graphical Diagram, (B) Cutaway
Diagram, (C) Pictorial Diagram, and (D) Combination Diagram.

Hydraulic circuits are usually drawn with symbols rather than with picture or cut-away-type drawings. Fortunately, hydraulic symbols are standardized throughout the industry. The graphical representation of the circuit (21-21A) is the usual way to show the hydraulic components and their relationships.

Hydraulics is a large and growing industry and can offer several advantages which make it an attractive power transmission system.

1. Hydraulics can produce linear or rotary motion without the use of gears, chains, or belts which may require more maintenance effort.

2. Hydraulic components lubricate themselves with hydraulic oil.

3. Hydraulic power is flexible — no shafts. A hydraulic hose can go around corners or flex with the machine motion or go across the room.

4. Hydraulics are capable of exerting forces to fit the need, tremendous forces or very delicate forces.

5. Hydraulic power is smooth and since hydraulic oil is virtually incompressible, instant power response, without jerks, is attained.

PNEUMATICS

Pneumatics and hydraulics are both fluid power systems and much that has been said about hydraulics will also apply to pneumatics. Pneumatics, of course, refers to the use of gases to transmit power from one point to another. The pneumatic system generates the pneumatic force, transmits, and controls the power it receives from the prime mover. Modern pneumatics goes beyond many of the historical applications of gases in motion such as the sailboat, windmills, and the like, because it involves generating and transmitting the gaseous flow within the pneumatic system itself. Whether or not the wind is blowing outside has no effect at all on modern pneumatics.

LAWS OF PNEUMATICS

But pneumatics is controlled by physical principles just as hydraulics is. Pascal's law, "the pressure at any point in a static fluid is the same in every direction and exerts equal force on equal areas," applies to both liquids and gases under pressure and allows pneumatics to gain a mechanical advantage just as hydraulics can.

The molecules of a gas are very far apart and very light. When a gas is heated, molecular activity increases in the gas, (this is true of solids and liquids too). Since there is more motion of the molecules, they tend to strike or bump into each other more often and if the gas is in a container, the molecules tend to strike the sides of the container more often and this is seen as an increase in pressure. Temperature has a more marked effect on the expansion of gases than it does on either liquids or solids. An example of the effect of temperature on a gas is an automobile tire that is soft in the morning but becomes hard after several hours of driving on the warm pavement.

MOLECULES VIBRATE

MOLECULES MOVE FREELY

RAPID MOVEMENT

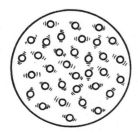
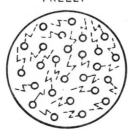
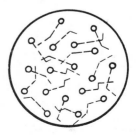

Fig. 21-22   Effect of Heat on Gas Molecules.

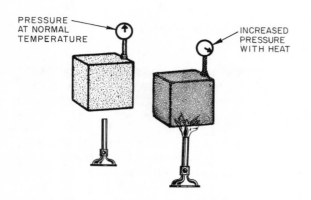

Fig. 21-23   Temperature Rise Affects Movement
and Pressure.

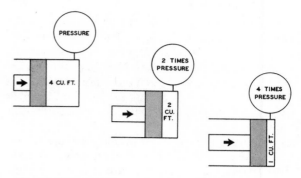

Fig. 21-24   Boyle's Law on the Volume of Gases.

The temperature behavior of gases was observed by Jacques Charles and his findings resulted in the physical law that bears his name. Charles' law states that all gases expand and contract in direct proportion to the change in the absolute temperature, provided the pressure is held constant, Fig. 21-23. V_1 and T_1 are original volume and temperature. V_2 and T_2 are final volume and temperature.

It follows that if the volume is held constant, then the pressure will vary in accordance with the temperature.

Boyle's law is one of the most common gas laws applied to pneumatics. Basically, this law states: "The absolute pressure of a confined body of gas varies inversely as the volume, provided the temperature remains constant." Double the pressure on a gas and its volume is reduced in half. Quadruple the pressure and the resulting volume is one-fourth the original volume, Fig. 21-24.

$$P_1 \times V_1 = P_2 \times V_2$$

P_1 and V_1 are the original pressure and volume. P_2 and V_2 are the pressure and volume after expansion or compression.

The formula can also be written as:

$$P_2 = \frac{P_1 \times V_1}{V_2} \quad \text{or} \quad V_2 = \frac{P_1 \times V_1}{P_2}$$

When these formulas are used in actual problems the pressures on the gages that are in the system must be converted to absolute pressure. Gages are calibrated at sea level pressure which is 14.7 p.s.i. The gage starting point or zero is therefore 14.7 less than absolute pressure of zero. Therefore, 14.7 p.s.i. is added to the original pressure, P_1, converting it to absolute pressure.

If P_2 is the unknown, 14.7 p.s.i. is subtracted from the answer to convert absolute to gage pressure.

$$P_1 \times V_1 = P_2 \times V_2$$

$$P_2 = \frac{P_1 \times V_1}{V_2}$$

$$P_2 = \frac{15 \text{ p.s.i.} + 14.7 \text{ p.s.i.} \times 5 \text{ cu.ft.}}{1 \text{ cu.ft.}}$$

$$P_2 = \frac{29.7 \times 5}{1} = 148.5 \text{ p.s.i.}$$

$$\begin{array}{r} \text{Absolute} = 148.5 \text{ p.s.i.} \\ - \ 14.7 \\ \hline 133.8 \text{ p.s.i.} \end{array}$$

$$P_2 = 133.8 \text{ p.s.i. Gage Pressure}$$

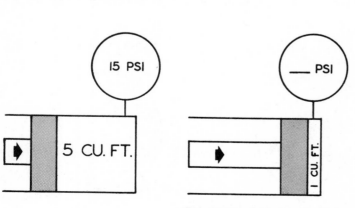

Fig. 21-25   A Problem in Volume and Pressure.

The same functions are required in a pneumatic system that are required in a hydraulic system. The main difference is that the transmitting media, a gas, is not rigid like a column of confined oil is. However, under pressure and in motion, gas can exert a force and produce work.

Air is the most common gas used, primarily because of its availability. Air has several desirable properties besides its abundance; it is nonpoisonous and nonflammable. However, several problems must be faced in using air as a fluid: (1) condensation may form in the system during temperature changes, (2) the oxygen in the air will support combustion, (3) the air may contain dirt that would be abrasive in the system if it were not carefully filtered out, and (4) air does not lubricate its own mechanisms as hydraulic oil does. Therefore, lubrication of components must be taken care of independently.

PNEUMATIC COMPONENTS

Physical laws govern the design of the parts that make up a pneumatic system which must have these components:

1. Pump or compressor to supply pressurized gas.

2. Receiver, storage cylinder, air cylinder, or air bottle to act as a reservoir for the pressurized gas.

3. Hoses or pipes to channel the air.

4. Valves to control pressure and direction of flow.

5. Motor which is acted on by the pneumatic fluid, converting pneumatic power back into mechanical power for use by the mechanism or machine.

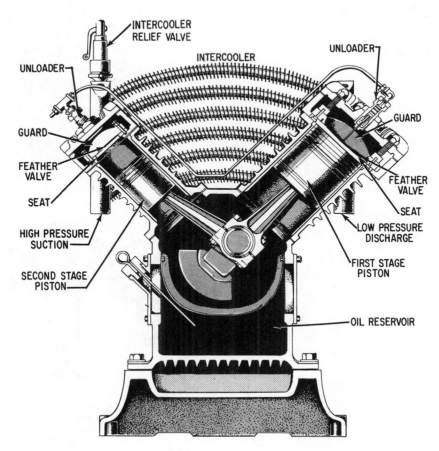

Fig. 21-26   A Simple Two-Stage Reciprocating Low-Pressure Air Compressor.

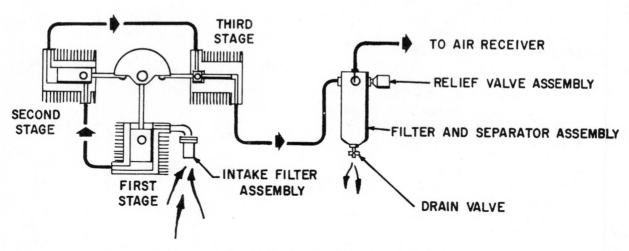

Fig. 21-27   Air Compressor Assembly.

Some of the components of a pneumatic system are identical to hydraulic components and nearly all are similar in their construction and operation; they are just designed to handle a different type of fluid.

Compressors

Air compressors are driven by an electric motor or by a prime mover. Compressors can be of centrifugal design, rotary design (gear, lobe, vane, rotary piston, etc.) or reciprocating types, Fig. 21-26. Air compressors often build up their final pressure over a series of stages. For example, in a three-stage reciprocating air compressor, Fig. 21-27, the air travels successively through three cylinders before it arrives at the air receiver or reservoir. The size of the cylinders and the piston displacement is progressively reduced as the volume decreases and the pressure increases. Special attention must be given to the lubrication of the many parts of all air compressors and many of the same techniques used in reciprocating internal combustion engines are used, such as splash oil systems and oil pumps.

Compressing air creates heat and a certain portion of the heat must be carried away to prevent damage to the compressor. Air cooling, water cooling, or oil cooling systems can be used.

Fig. 21-28   A Piston Poppet Pressure Actuated Air Valve.

Fig. 21-29   A Three-Way Unit for Filtering Air, Lubricating Air, and Regulating the Air Pressure.

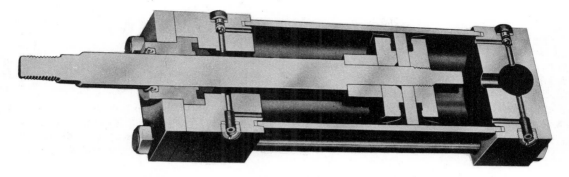

Fig. 21-30   Air Cylinders are very Similar to Hydraulic Cylinders.

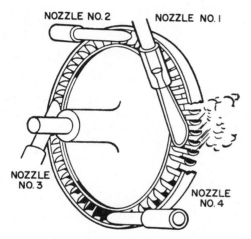

Fig. 21-31   Single-Stage Turbine.

Air Cylinders

Motors use the compressed gas to apply force to the machine and its work load. Air cylinders are very similar to hydraulic cylinders, Fig. 21-30. They can be single-acting, double-acting, or telescoping cylinders. The double-acting cylinder is the most common — the fluid can be directed to either side of the piston giving power alternately in each direction.

Turbine Motors

If rotary motion is the desired output of the pneumatic system, a turbine motor is usually employed. Jets of air are directed against the turbine blades. The construction principles are much the same as those found in steam turbines or gas turbines but usually the application is very much smaller and lighter weight. Both single stage and multiple stage turbines are used, Fig. 21-31 and 21-32.

Another turbine is the pinwheel turbine which is pure and simple jet action applied to pneumatics. Its action is like that of a lawn sprinkler, Fig. 21-33.

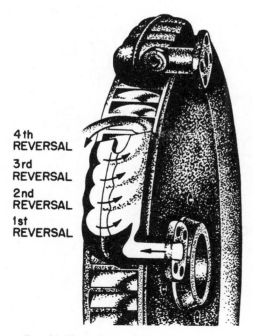

Fig. 21-32   Multiple-Stage Air Turbine.

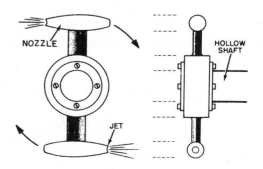

Fig. 21-33   Pinwheel Air Turbine.

SUMMARY

Pneumatics has many applications and in a sense it competes with hydraulics, but both are part of the fluid power team. Most persons could identify some of the many pneumatic applications, such as air hammers or compressed air tools used in manufacturing operations but these are only a few of the many modern applications.

1. Tools such as grinders, buffers, sanders, drills, screwdrivers, nut setters, and wrenches.
2. Air hoists.
3. Rivet, chipping, and sand hammers.
4. Clamping devices.
5. Blast cleaning.
6. Spraying.
7. Air control circuits.
8. Pneumatic systems as backup or emergency systems for hydraulic systems.
9. Air gaging.
10. Pneumatic conveying of materials.

Another phase of fluid power involves vacuum applications. Vacuum is just the opposite of fluid under pressure. It is fluid (usually air) removed from a space. Vacuum applications depend on atmospheric pressure. Simple vacuum principles can be seen in a collapsing soda straw — reducing pressure on the inside of the straw allows the atmospheric pressure, 14.7 p.s.i., to push in and collapse the straw.

Atmospheric pressure working against areas of very low pressure or partial vacuum can be used to clamp and lift materials, Fig. 21-34, for bonding and laminating materials. Vacuum cups can "lock" on to any smooth surface.

The vacuum forming of pliable plastic sheets is another application. A vacuum pump produces the low pressures that are necessary to put this principle to work, Fig. 21-35.

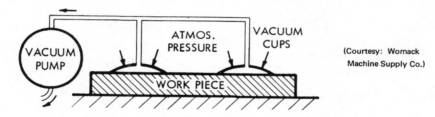

(Courtesy: Womack Machine Supply Co.)

Fig. 21-34   Lifting Concrete by Vacuum Clamping.

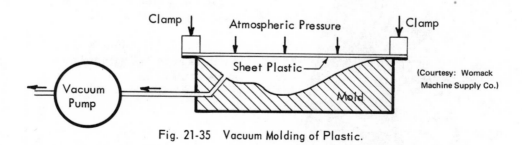

(Courtesy: Womack Machine Supply Co.)

Fig. 21-35   Vacuum Molding of Plastic.

GENERAL STUDY QUESTIONS

1. Explain several ways in which fluid power has been used throughout history.

2. Define the term hydraulics.

3. Define the term pneumatics.

4. What can be said about the compressibility of liquids?

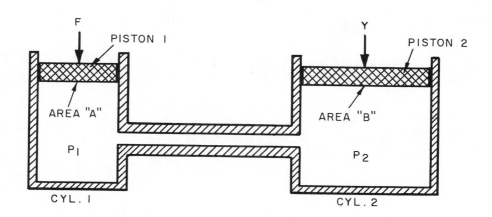

5. The hydrostatic principle tells us that $P_1 = P_2 = F/A = Y/B$ (assuming no pressure deviation). If A = 1 sq. in., F = 1 lb., and B = 10 sq. in., what is "Y"?

6. How can you compare the mechanical advantage possibilities of hydraulics to that of a lever?

7. What can cause excessive friction on fluids flowing in hoses or pipes?

8. What effect does turbulence have on the efficiency of fluid dynamics?

9. What does Bernoulli's principle deal with?

10. List several common, everyday applications for hydraulic systems.

11. List the basic components of a typical hydraulic system.

12. Why are filters of particular importance to a hydraulic system?

13. What is the basic difference between a positive displacement pump and a non-positive displacement pump?

14. What three purposes can valves be designed to satisfy?

15. What hydraulic device is used to provide rotary hydraulic power at the load?

16. List several applications for hydraulic cylinders.

17. List several advantages that the field of hydraulics can offer.

18. Why does an increase in temperature have more of an effect on the pressure within a pneumatic system than it does in a hydraulic system?

19. State Boyle's law in a brief, basic form.

20. What basic components are necessary for a pneumatic system?

21. Prepare a list of several modern uses for pneumatic devices.

CLASS DISCUSSION TOPICS

● Discuss the term "fluid."

● Discuss how both hydraulic and pneumatic systems are systems for the transmission of power and not prime movers themselves.

● Discuss the similarities that exist between hydraulics and pneumatics.

● Discuss the term "dynamics" as it relates to fluids.

● Discuss how manufacturing technology could play a part in the development of hydraulics.

● Discuss why high pressures are often needed in a hydraulic system.

● Trace the path of hydraulic fluid through a spool valve assembly such as that shown in Fig. 21-17.

● Discuss how the linear motion of a hydraulic cylinder could be converted into rotary motion.

● Discuss the several physical laws that apply to pneumatics.

Unit 22

ATOMIC ENERGY

The era of atomic energy is still in its infancy but this energy source is by far the most promising, the most awe-inspiring, and the most fearful that we have yet uncovered. The tremendous energy reserve of atomic power promises to be an ecomonical power source for all nations for centuries to come. Atomic power can be relied on long after conventional power reserves of coal and oil are exhausted. Of course, when atomic energy is used natural resources of coal and oil can be conserved.

It is difficult for most people to comprehend the energy locked within the atom: it is almost inconceivable that enormous energy exists within something that small. Yet, the potential energy that lies within a pound of uranium (atomic fuel) is nearly equal to 3,000,000 pounds of coal, twenty-five railroad cars.

No one in today's world is without a fear of the destructive possibilities of atomic and nuclear power. The strongest nations of the earth are armed and poised with atomic weapons that are capable of destroying our entire civilization. A single nuclear warhead could wipe out an entire city and still have its effects felt hundreds of miles away. The shadow of atomic destruction covers all. When nations can learn to reduce world tensions and the resulting atomic arms struggle, the shadow of atomic destruction will be lifted to expose the brightness of peaceful atomic progress.

An appreciation of atomic energy and an understanding of how it is produced and controlled is a valuable accomplishment for all persons. Basic knowledge of the structure of matter should be gained, however, before a more detailed investigation of atomic energy is begun.

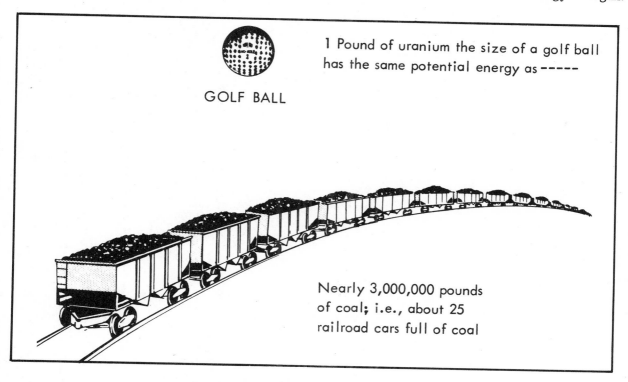

GOLF BALL

1 Pound of uranium the size of a golf ball has the same potential energy as -----

Nearly 3,000,000 pounds of coal; i.e., about 25 railroad cars full of coal

Fig. 22-1 Some Comparisons of Magnitude.

NATURE OF MATTER

The smallest complete unit of matter is called the atom. Atoms are the building blocks of all substances. If it were possible to magnify any substance an infinite number of times, it would be seen as being made up of atoms. Some substances are made up of all the same type of atoms; these substances are the elements. There are ninety-two different elements (types of atoms) that normally occur in nature (additional elements have been man-made). A cluster of atoms makes up a molecule.

Other substances could be "magnified" and be found to be made up of different kinds of atoms that are bound together. A cluster of different atoms also makes up a molecule of a substance. However, when two or more different atoms make up a molecule it is referred to as a compound, not an element. There are many ways that atoms can arrange themselves; many different compounds can be formed. Over 400,000 different combinations are known (wood, cloth, water, coal, rock, etc.). All of these substances are made of the various combinations of the basic elements.

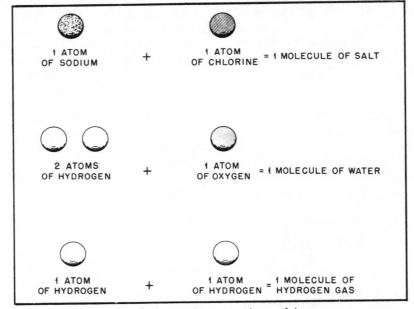

Fig. 22-2   Molecules are made up of Atoms.

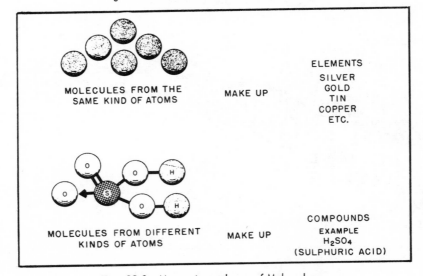

Fig. 22-3   Matter is made up of Molecules.

The atom itself is made up of protons, electrons, and with the exception of hydrogen, neutrons. The protons are heavy particles at the center of the atom: they have a positive electrical charge. The neutrons are also heavy particles at the center of the atom; they have no electrical charge. The protons and neutrons, bound tightly together, form the center or nucleus of the atom. The electrons are in orbit around the nucleus: they are extremely light and carry a negative electrical charge. The atom might be thought of as a solar system in miniature. Electrons are held in their orbits by the pull of the protons in the nucleus. The number of electrons in orbit around the nucleus is equal to the number of protons in the atom.

The electrons are arranged in neat and predictable patterns or shells around the nucleus of the atom. Some elements are chemically satisfied, all of their shells are complete. They do not react with other elements, they are inert. Most elements have atoms with incomplete outer shells. The electrons in these outer shells are loosely held and they can, in a sense, combine with other atoms. These atoms can share the electrons of their incomplete outer shell with those of other atoms to form a molecule. This sharing of electrons produces chemical compounds. An example would be sodium chloride (common table salt). The chlorine atom has only seven electrons in its outer shell, it is one short. The sodium atom has only one electron in its outer shell. These two atoms share their outer shell electrons, each outer shell is now complete, and a stable compound is formed.

Common table salt is known chemically as sodium chloride (NaCl) since one atom of sodium combines with one atom of chlorine to form a molecule of salt.

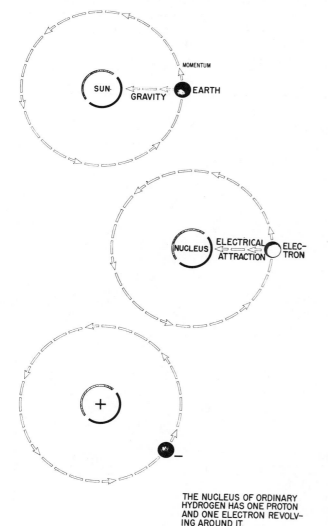

THE NUCLEUS OF ORDINARY HYDROGEN HAS ONE PROTON AND ONE ELECTRON REVOLV-ING AROUND IT.

Fig. 22-4 An Atom of Hydrogen.

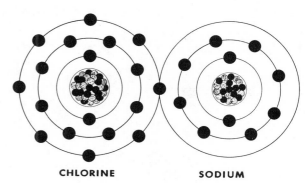

CHLORINE   SODIUM

Fig. 22-5 Molecule of Sodium Chloride (NaCl).

Chlorine normally has only 7 electrons in its outermost range or shell. For an atom to be chemically inert, or satisfied, it should have 8 electrons in this shell. Sodium has only one electron in its outermost shell. Now, if one atom of sodium combines with one atom of chlorine in such a way that the single atom in the sodium shell is shared with the outer shell of chlorine, then each atom has 8 electrons in its outermost shell. Under these conditions, a stable chemical compound results from the chemical reaction between sodium and chlorine and a molecule of sodium chloride is formed.

There is space between the molecules of any substance; some molecules are very close together, some very far apart. Gases have the most space between molecules. Liquids have less space between their molecules. Molecules of solids have the least space between them; they are the most dense.

All molecules are in constant motion. In gases there is tremendous movement of the molecules while in solids the movement is reduced to a slight vibration. Also, the molecule's speed increases as it is heated. This can be seen in metals; they expand when they are heated. It can also be seen when water boils. Some molecules get so hot and move so fast that they are thrown off the surface and appear as steam.

The atom of each element has an <u>atomic weight</u>. This weight is the weight of the nucleus, combined number of protons and neutrons. Hydrogen, the lightest element, has one proton; its atomic weight is 1. Zinc has 30 protons and 35 neutrons; its atomic weight is 65. Uranium, a very heavy element, has 92 protons and 146 neutrons; its atomic weight is 238. The electrons are infinitely light and are not considered in atomic weight. Elements are also assigned <u>atomic numbers</u>, the number of protons in the nucleus.

Another point in the understanding of atomic energy is isotopes. <u>Isotopes</u> are forms of the same element that have different atomic weights. The isotopes of an element are chemically alike and have the same atomic number (number of protons) yet their total weight is different. The difference is in the number of neutrons in the nucleus. Isotopes are common; all elements have them. There are more than 1,000 isotopes known today.

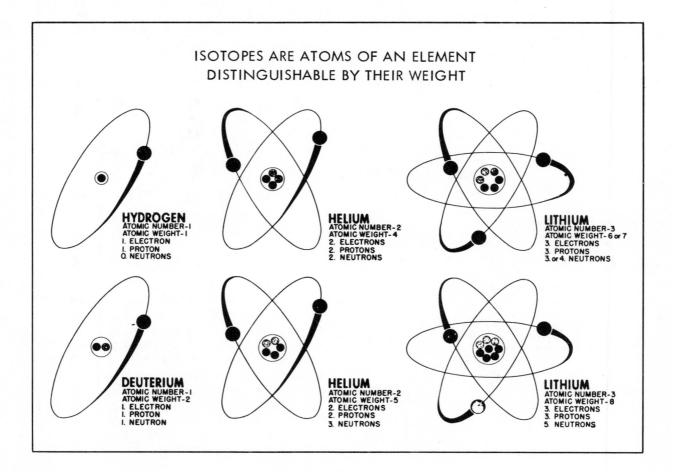

ISOTOPES ARE ATOMS OF AN ELEMENT
DISTINGUISHABLE BY THEIR WEIGHT

HYDROGEN
ATOMIC NUMBER-1
ATOMIC WEIGHT-1
1. ELECTRON
1. PROTON
0. NEUTRONS

HELIUM
ATOMIC NUMBER-2
ATOMIC WEIGHT-4
2. ELECTRONS
2. PROTONS
2. NEUTRONS

LITHIUM
ATOMIC NUMBER-3
ATOMIC WEIGHT-6 or 7
3. ELECTRONS
3. PROTONS
3. or 4. NEUTRONS

DEUTERIUM
ATOMIC NUMBER-1
ATOMIC WEIGHT-2
1. ELECTRON
1. PROTON
1. NEUTRON

HELIUM
ATOMIC NUMBER-2
ATOMIC WEIGHT-5
2. ELECTRONS
2. PROTONS
3. NEUTRONS

LITHIUM
ATOMIC NUMBER-3
ATOMIC WEIGHT-8
3. ELECTRONS
3. PROTONS
5. NEUTRONS

Fig. 22-6 Isotopes.

Some isotopes that occur in nature are radioactive, they emit rays or radiation. Radiation is not new; we have always been exposed to radiation. The stars are intensely radioactive and the earth is slightly radioactive. There are many types of radiation: light radiation which we can see, heat radiation which can be felt, X-radiation which can neither be seen nor felt, and there is radiation from radioactive material. Our senses cannot dedect radiation from radioactive material. Until recently, radiation from radioactive material occurred only in nature; now there are many artificially made radioactive materials.

Radiation is a stream of fast-flying particles or waves coming from atoms. There are three types of radiation: they are, (1) alpha, (2) beta, and (3) gamma. The <u>alpha</u> particle is heavy and travels a short distance and can be stopped by human skin or a piece of paper. The <u>beta</u> particle is the same as an electron; it, too, can be easily shielded. The <u>gamma</u> ray is the most penetrating and dangerous. It is the radiant energy from the excess energy in the atom's mass. Great care must be taken to protect living tissue from gamma rays. Radiation, like fire or electricity, can be dangerous; it is a matter of the degree of exposure.

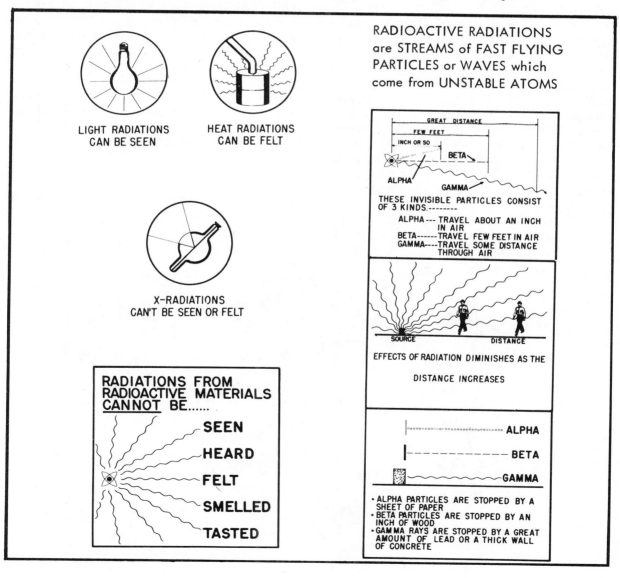

Fig. 22-7   Radiation.

Uranium is a naturally radioactive element. Down through the centuries the uranium atom loses protons in the form of radiation. The uranium slowly becomes radium and then finally becomes lead, a stable element. This process of changing one element into another is called natural transmutation.

With an understanding of the nature of matter, atomic structure, and radiation, further study of atomic energy can follow.

ATOMIC FISSION

As scientists studied the atom, they found that the atomic weight of each atom was less than the weight of the total mass of protons, neutrons, and electrons that it is composed of. "The whole is less than the sum of its parts." Some of the mass appeared to have disappeared, contrary to the Law of the Conservation of Matter, "matter can be neither created nor destroyed, but only altered in form." As early as 1905, Albert Einstein reasoned that mass and energy were two forms of the same thing. He surmised that the loss in mass was converted into energy. When free protons and neutrons were formed into a nucleus, they lost some mass, (packing loss). This loss in mass was converted into the energy that binds the nucleus together, (binding energy). If the atom could be broken or split, this binding energy could be released. Einstein calculated that this energy release would be immense.

Uranium-235 has been found to be the only natural element that can be split. An atom of uranium-235 can be split when its nucleus is struck by a free neutron that is traveling at the correct speed. A neutron "bullet" can strike the nucleus of the atom and split it.

Suppose that free neutrons are being shot (like a machine gun) at a mass of uranium atoms. Many of the neutrons will miss their target completely. However, when a neutron does hit the nucleus of the uranium atom, the atom will split. The result of this split is two smaller atoms, krypton (90) and barium (143). Also two neutrons are freed. The process of splitting the uranium-235 atom is called fission. Moreover, just as Einstein had predicted, tremendous energy is released during the process.

Other isotopes of uranium, U-234 and U-238, will not undergo fission. Unfortunately, 99.3 percent of uranium occurs as U-238, only 0.7 percent is the valuable U-235. The amount of U-234 is too small to be significant.

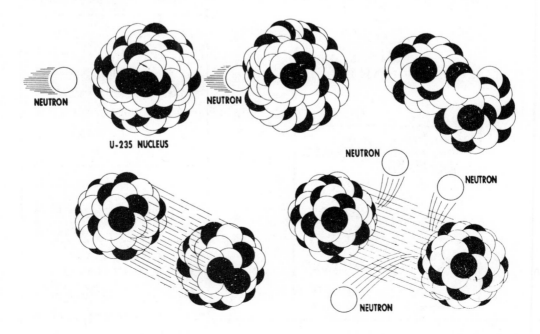

Fig. 22-8 Fissioning of Uranium - 235.

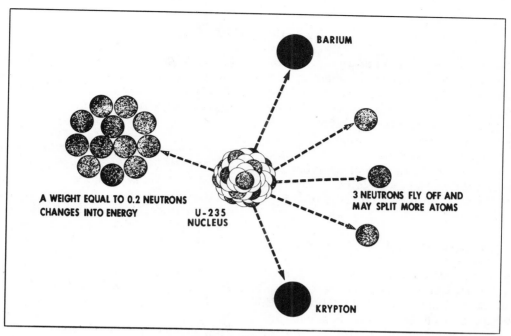

Fig. 22-9   Weight (Mass) Into Energy.

A chain reaction is begun with the initial fission of the first uranium-235 atom. The free neutrons released in the fission travel until they strike and split other uranium atoms. With each fission in the chain reaction, energy is released and more neutrons are set free to strike and split still more atoms. An almost unlimited number of free neutrons can be released in an incredibly short amount of time. The result is an enormous release of energy. If the chain reaction is left uncontrolled, we have an atomic bomb, a monstrously destructive release of energy. The chain reaction can be controlled, however, with the use of an atomic reactor.

A chain reaction cannot be supported by a single atom. Neither can the reaction be supported by only two, three, or four atoms. There must be a large enough mass of uranium atoms present to support and propagate a chain reaction. If the mass is too small, the reaction will die out because too many neutrons are escaping from the surfaces. A mass where neutrons escape from the surfaces rapidly is said to be "sub-critical". In order to maintain a productive chain reaction, there must be more than one additional atom split for each atom previously split. When a productive chain reaction is maintained, the mass is said to be

"super-critical". When the process of nuclear fission is proceeding at a controlled rate, as in a nuclear reactor, the mass is said to be simply "critical".

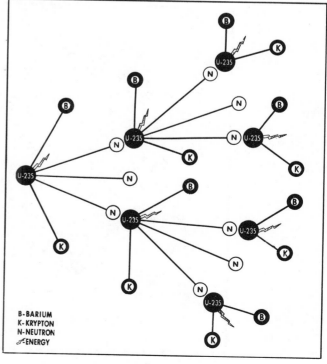

B- BARIUM
K- KRYPTON
N- NEUTRON
⚡-ENERGY

Fig. 22-10   Chain Reaction.

241

NUCLEAR REACTOR

The nuclear reactor is designed to provide a controlled fission of the uranium atoms, releasing energy at a constant rate for a long period of time. The chain reaction is slowed down by diluting the uranium with another substance such as graphite. As fission occurs, the free neutrons are slowed down by the graphite moderator. This combination of uranium and graphite is referred to as a "pile". The first atomic pile was built at the University of Chicago in 1942.

In Fig. 22-11 the initiating neutron works its way through the graphite (carbon) atoms, striking a uranium nucleus at A, and causing its fission. Two more neutrons are emitted. The neutron following path (1) collides with graphite nuclei. These collisions reduce its speed until it causes the disintegration of another uranium nucleus at B. The disintegrating nucleus gives two more neutrons. The neutron following path (3) passes through the uranium mass C without causing disintegration, and proceeds to D, where it causes further reaction. The neutron proceeding along path (2) is captured by the uranium nucleus at E, since it is moving at a relatively high speed. The neutron following path (4) escapes through the surface of the pile.

The neutrons released by the fission of the U-235 may do several things. First, if the chain reaction is to be maintained, at least one of the neutrons must strike the nucleus of another U-235 atom causing its fission. Second, some of the neutrons that have been slowed down may be captured by the U-238 isotope. There are 140 of this heavier uranium isotope for every one U-235 isotope. When U-238 captures a neutron it becomes U-239. Neptunium U-239, which is radioactive, soon decays into a more stable form of U-239 - Plutonium. The formation of plutonium is important because it, too, is fissionable. Plutonium then can be used to help sustain the chain reaction or it can be left in the reactor for recovery at a later date. Third, the neutron can escape the pile. This is undesirable and should be kept to a minimum since it represents a loss.

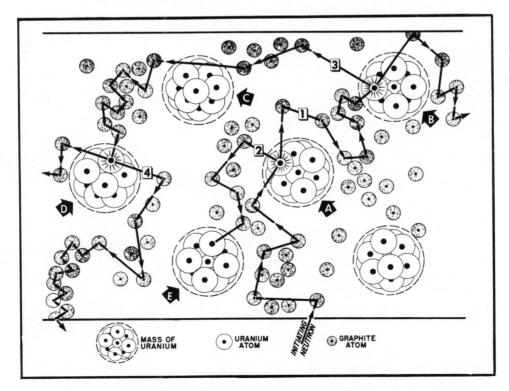

Fig. 22-11   Action of Uranium-Graphite Pile.

The nuclear reactor itself is made up of several main parts. Atomic fuel, uranium U-235 is distributed within tons of U-238. The moderator, usually carbon in the form of graphite, or heavy water, has the effect of slowing the neutrons and also slowing the chain reaction. The control and safety rod mechanism can regulate the chain reaction properly. Boron and cadmium rods can be moved in and out of the atomic pile to control fission. Both of these elements have a strong ability to capture neutrons. When they are well into the pile they can capture so many neutrons that there will not be enough left to sustain the chain reaction. If the rods were all the way out of the pile, the chain reaction could reach sufficient intensity to burn up or melt down the materials. The rods are positioned to allow the desired level of reaction. The shield of a reactor is important for the protection of operating personnel. Concrete, several feet thick, and iron are used to completely enclose the reactor. The fission process emits neutrons and gamma rays that are very dangerous. The shield traps these dangerous rays and particles. The coolant is necessary to keep temperatures in the reactor low enough to prevent heat damage. The coolant may be water or other fluids. In power reactors, the coolant picks up this heat energy and uses it to produce usable power.

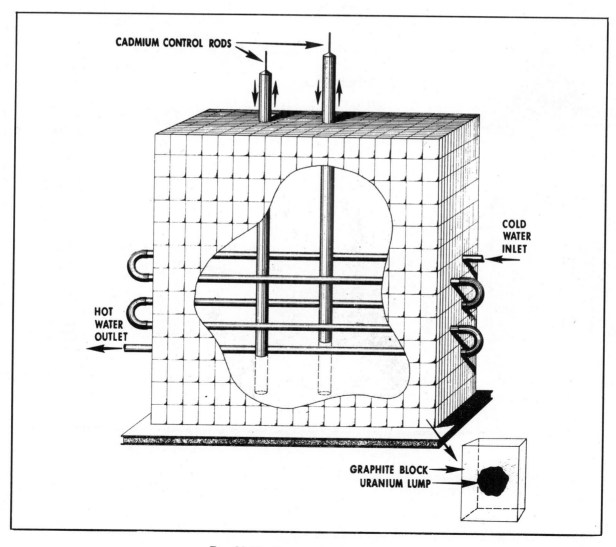

Fig. 22-12   Uranium-Graphite Pile.

Perhaps the greatest potential use for the nuclear reactor is the production of electricity. The intense heat energy of the reactor can be converted into electrical power. The action of a nuclear reactor which is producing electricity is, in many respects, similar to that of a coal or oil-fired electrical generating station.

The intense heat energy inside the reactor is picked up by the coolant or heat exchange medium that flows in pipes within the pile. When heat is transferred to ordinary water in a heat exchanger or boiler, high-pressure steam is produced. The high-pressure steam is directed against the turbine blades, revolving the turbine which drives the electrical generator.

Many nuclear reactors, designed for electrical generation, are being built throughout the nation. There are many other uses for atomic reactors: atomic-powered submarines, atomic-powered merchant vessels. Atomic locomotives and airplanes are a possibility of the future.

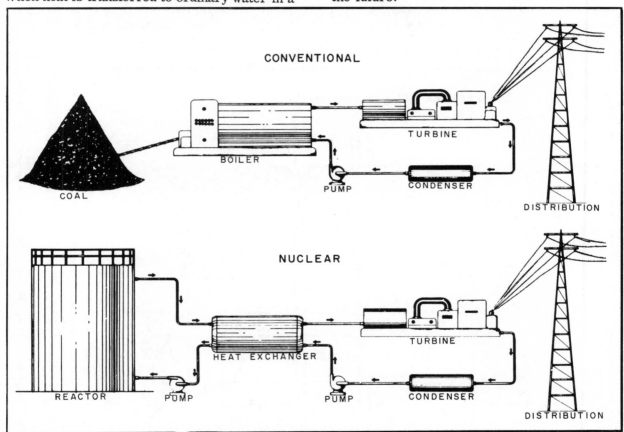

Fig. 22-13 Electrical Power Plants.

GENERAL STUDY QUESTIONS

1. Define the following terms: a. Atom, b. Element, c. Molecule, d. Compound

2. What parts make up an atom? Explain their location in the atom.

3. What happens when atoms share their electrons?

4. Define atomic weight.

5. Define atomic number.

6. Define isotopes.

7. What is radiation? What are the three types of radiation?

8. What is fission?

9. List the products of uranium-235 fission.

10. Explain how a chain reaction takes place.

11. What is a nuclear reactor designed to accomplish?

12. What is plutonium?

13. List the main parts of a nuclear reactor and briefly explain the function of each part.

14. Explain how the heat energy within a nuclear reactor can be converted into electrical energy.

CLASS DISCUSSION TOPICS

● Discuss the destructive power of atomic weapons.

● Discuss the peaceful uses of atomic energy.

● Discuss the nature of matter - atoms, elements, molecules, compounds.

● Discuss the theory of atomic structure.

● Discuss radioactive materials and explain why shielding and safety precautions are necessary for these materials.

● Discuss why energy is released during fission.

● Discuss chain reaction, sub-critical mass, super-critical mass, and critical mass.

● Discuss the operation of a nuclear reactor.

Unit 23

THE DIRECT CONVERSION OF ENERGY

Energy should be put to work in as direct a manner as is possible. The long route of changing energy through several different states is basically costly and wasteful. Under usual circumstances energy cannot be destroyed, but it can surely be lost to friction or be converted into other forms that are not really involved in producing the desired power.

The energy conversion matrix below illustrates this very well. Let's follow through the energy transformation involved in producing electricity which is the most universally used type of power for home and industry. The chemical energy of coal is converted into thermal energy as it burns. This thermal energy is absorbed by water; steam is produced and directed against the blades of a turbine, converting the thermal energy into kinetic energy. The kinetic energy of the turbine, in turn, rotates the generator and the energy is at last converted into electrical energy for transmission and use. The electrical energy may now again take many forms as it is used by the consumer. The described method is good, it is reliable, it produces the vast portion of all electrical energy but it does represent the "long route" of energy conversion.

TO ⬇ FROM ➤	ELECTROMAGNETIC	CHEMICAL	NUCLEAR	THERMAL	KINETIC (MECHANICAL)	ELECTRICAL	GRAVITATIONAL
ELECTRO-MAGNETIC		Chemiluminescence (fireflies)	Gamma reactions (Co^{60} source) A-bomb	Thermal radiation (hot iron)	Accelerating charge (cyclotron) Phosphor	Electromagnetic radiation (TV transmitter) Electroluminescence	Unknown
CHEMICAL	Photosynthesis (plants) Photochemistry (photographic film)		Radiation catalysis (hydrazine plant) Ionization (cloud chamber)	Boiling (water/steam) Dissociation	Dissociation by radiolysis	Electrolysis (production of aluminum)	Unknown
NUCLEAR	Gamma-neutron reactions ($Be^9 + \gamma \rightarrow Be^8 + n$)	Unknown		Unknown	Unknown	Unknown	Unknown
THERMAL	Solar absorber (hot sidewalk)	Combustion (fire)	Fission (fuel element) Fusion		Friction (brake shoes)	Resistance-heating (electric stove)	Unknown
KINETIC	Radiometer Solar cell	Muscle	Radioactivity (alpha particles) A-bomb	Thermal expansion (turbines) Internal combustion (engines)		Motors Electrostriction (sonar transmitter)	Falling objects
ELECTRICAL	Photoelectricity (light meter) Radio antenna Solar cell	Fuel cell Batteries	Nuclear battery	Thermoelectricity Thermionics Thermomagnetism Ferroelectricity	MHD † Conventional generator		Unknown
GRAVITATIONAL	Unknown	Unknown	Unknown	Unknown	Rising objects (rockets)	Unknown	

† Magnetohydrodynamics.

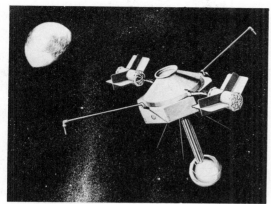

A. Atomic generators convert heat from radioactive fuel into electricity to power radio transmitters of satellites.

B. A nuclear-powered device (center) supplied electrical power for explorations on the moon.

C. Atomic fuel provides the power for instrumentation of this weather satellite.

D. A nuclear-powered laboratory for soft landing on the lunar surface.

Fig. 23-1 U.S. Atomic Energy SNAP program (Systems for Nuclear Auxiliary Power) Utilizes Direct Energy Converters.

How much better it would be if an energy form could be converted directly into electrical energy. The idea is not impossible and it is certainly not new. Several direct conversion techniques have been demonstrated by scientists in the laboratory for many years. And the conventional battery has also been converting chemical energy directly into electrical energy for a good long while, but they do have limitations of weight and length of service.

In our space age technology, many of the laboratory phenomena of direct energy conversion have been and continue to be under intensive development. The result has been a whole new family of direct energy conversion devices. A good deal of the impetus for the development of these energy conversion devices has come from the nation's space program.

Space systems need rockets for basic propulsion, but in addition they need electrical systems that fit the requirements of long service life, low weight, and high reliability. Direct energy converters can meet these requirements and they play their part in meeting the electrical/electronic and life support functions involved in outerspace projects.

The direct energy conversion devices have applications beyond their uses in space exploration wherever long life, low weight, and high reliability are of great importance. These energy converters are now the heart of several rapidly maturing industries. In many cases, these products are out of the laboratory, off the drawing board, and are being manufactured and put into space, military, and commercial use.

When direct energy converters are discussed, the topic really boils down to the various devices that are capable of producing electricity directly from one of the primary energy sources. Several direct energy converters will be discussed in this unit.

Solar Cell - Electrical energy from the sun; electromagnetic energy.

Thermophotovoltaic (TPV) Converter - Electrical energy from thermal photons.

Fuel Cell - Electrical energy from chemical energy.

Nuclear Battery - Electrical energy from nuclear energy.

Thermoelectricity and Thermionics - Electrical energy from thermal energy.

Magnetohydrodynamic (MHD) Generator - Electrical energy from kinetic energy.

These energy converters do not rely on the dynamic conversion of energy through reciprocating or rotary machines -- quietness of operation is inherent in these energy converters since there are no moving parts. These techniques are highly sophisticated in their manufacture, but still very reliable and safe under operating conditions.

Does science and research apply to the study of these energy converters? Yes, of course it does. While heat engines are basically the domain of the mechanical engineer, direct energy converters are more the domain of the chemist, physicist, and nuclear scientist.

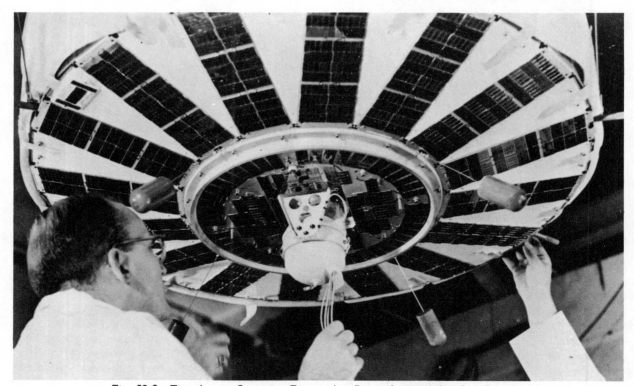

Fig. 23-2   Tiny Atomic Generator Fastened to Base of a TRANSIT Satellite. In the orbiting generator, atomic fuel generates heat which is converted through thermocouples to electrical energy.

SOLAR ENERGY

The sun gives our earth heat and light and, therefore, it gives us the gift of life itself. Without the sun, life as we know it would cease to exist. The huge, intensely hot body of gases that we call the sun, is also responsible for most of the power that we harness or liberate for our benefit.

The water cycle, warm air picking up water vapor, cooling, condensing, and falling to the earth again in the form of rain, fills our rivers and streams. We harness the energy of flowing water with hydroelectric plants to produce electricity.

The sun makes the process of photosynthesis possible. Green plants use the sun's energy to grow; our bodies benefit as we eat the plant's stored energy. Photosynthesis, taking place over the past ages, has made possible large deposits of coal and petroleum, our primary power sources. When we burn coal to heat homes, to generate electricity, or to carry on industrial processes we are really releasing the sun's energy that has been stored for thousands of years. In a similar manner, petroleum products are burned to produce power.

The wind is another source of energy that the sun is responsible for. For centuries the air currents that travel across the earth's surface have driven ships. The harnessing of wind to mills and pumps of various types has also been carried on for many centuries.

Most of the sun's energy goes unused. When compared to the total amount of sun energy that falls on our earth, the energy of flowing water, coal, petroleum, natural gas, and wind are weak. The sun showers the earth with one-thousand trillion-plus kilowatt hours of energy every day and only a very small fraction of the energy has ever been directly converted for use. We have to wait for plants to grow or for rain to fall in order to get the sun's energy.

For centuries the idea of using the power of the sun directly has intrigued scientists.

Early devices used reflectors and mirrors to concentrate the heat of the sun's rays. The concentrated heat was usually applied to the boiler of a steam engine, the engine being connected to a pump, generator or some other machine. Work and experimentation along these lines is still carried on. The sunlight is free but there is considerable expense involved in the initial installation and in the maintenance of such systems.

Other devices that directly convert the sun's radiant energy into electrical power have been in existence for many years. The thermopile had its beginnings in 1823, the photogalvanic cell can be traced to the year 1839, and the barrier-layer photovoltaic cell was developed around 1876. These devices worked and still have many applications today but they do not produce much electricity. They are, therefore, suitable for measuring instruments and controlling devices that require small voltages. The three devices mentioned are generally unsuitable for producing electricity in quantity or in competition with conventional electric generators. At best, these units are only one percent efficient when converting solar energy into electrical power.

The thermopile consists of a circuit that contains two dissimilar conducting materials. When one junction of the conductors is held at a different temperature than the other junction of the conductors, an electromotive force is set up between the junctions. Thermopiles (a group) and thermocouples (a single unit) are often used to operate relays and electronic controls.

The photogalvanic cell is made of two electrodes suspended in an electrolyte. When light falls on only one of the two electrodes, an electromotive force is produced.

The photovoltaic cell produces an electromotive force when it is exposed to light. A light-sensitive semiconductor such as selenium is used in this cell. The common photocell is often used in photographic exposure meters, photoelectric eyes, and photoswitches.

No significant breakthroughs in the search for ways to directly convert more of the sun's radiant energy into electricity were recorded until 1954. In that year three scientists from the Bell Telephone Laboratories reported that they had improved the photovoltaic cell and increased its efficiency to six percent. The scientists, D. M. Chapin, C. S. Fuller, and G. L. Pearson, later succeeded in raising the efficiency of their silicon photovoltaic cell to eleven percent. A device with an efficiency of eleven percent does have possiblities of becoming a practical electrical power source. To cite an illustration, on a clear day in Phoenix, Arizona, if 1000 watts of solar energy falls on every square meter of ground, 110 watts of electrical power could be produced by a square meter of the silicon photovoltaic cell.

Even though intense research continues toward improving the photovoltaic cell (commonly called the solar battery), the device has already found an important use in powering communications satellites that can relay conversations, radio, and television around the world. The communication satellite Telstar is powered by 3600 solar batteries. Solar batteries have been successfully used in experiments in which they powered rural telephone systems. Without a doubt, many uses for solar batteries will be found in the near future. Costs of production are now very high. The device is now largely experimental but someday it will be as commonplace as present day electrical sources.

Launched July 19, 1962, the Telstar satellite was assembled in a surgically clean environment at Bell Telephone Laboratories in Hillside, N. J. Dark squares on the surface of the satellite are solar batteries that convert sunlight into electrical energy needed to power Telstar's electronic package. Bands running around the waist of the satellite are receiving and transmitting antennas.

Fig. 23-3 Prototype Model of Telstar.

THE SOLAR BATTERY

First of all, the solar battery is made of semiconducting materials. We are familiar with electrical conductors and electrical insulators. Conductors are materials with loosely bound electrons that can easily move about. Insulators are materials with tightly bound electrons that cannot move about easily. Ninety percent of the substances we know are either conductors or insulators. However, there are a few substances that cannot be classed as either conductors or insulators: they are called semiconductors. Silicon and germanium are examples of semiconductors.

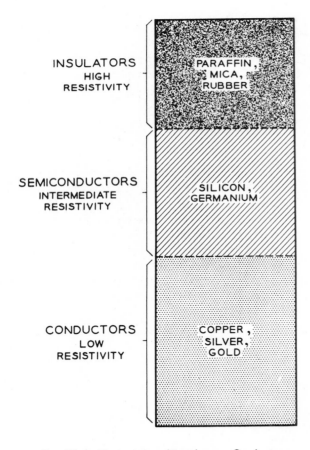

Fig. 23-4 Resistivity of Insulators, Conductors, and Semiconductors.

Insulators have a high resistivity to the flow of electrons while conductors have a low resistivity to the flow of electrons. Semiconductors lie in the middle ground and cannot be classed as either conductors or insulators.

The conducting qualities of semiconductors will change with the temperature of the material. As a semiconductor is heated it becomes a better conductor of electricity; more electrons are free to leave their atoms and move about. The reverse is also true: cooling the semiconductor makes it a better insulator, fewer electrons can break out of their orbits and become free electrons.

In Fig. 23-5 we can see what takes place in a semiconductor. Though oversimplified, the drawings do illustrate the point. In (A) we see a tray with ball bearings covering its bottom; the ball bearings represent the semiconducting material. Drawing (B) illustrates the semiconductor at absolute zero. The tray is tipped (electrical field) but the electrons will not flow and so the material will not conduct. In (C) the semiconductor is heated, the tray is shaken (thermal energy) and an electron hops out of orbit. The material becomes conductive. In (D) a potential is again applied; the free electrons can now move. The ball bearing moves to the lower end leaving a "hole" at the upper end.

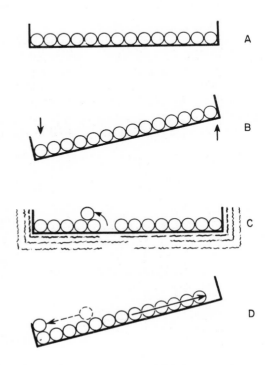

Fig. 23-5 Ball Bearings in a Tray Used to Illustrate the Action of a Semiconductor.

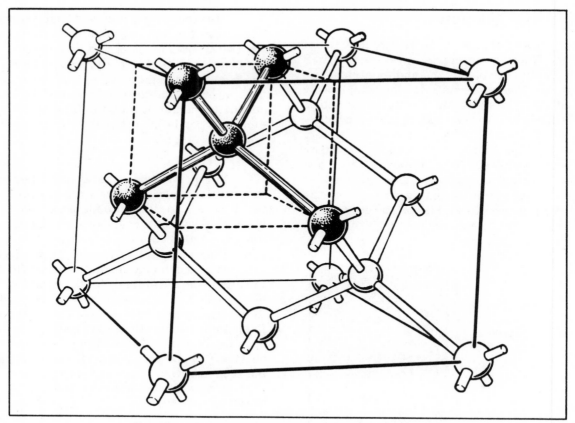

Fig. 23-6   Crystal Lattice Structure of Silicon or Germanium.

Semiconductors are heat sensitive and photosensitive. Light falling on a semiconducting material causes more electrons to leave their orbits, creating free electrons and electron short holes. Photons are light energy and their effect on a semiconductor could be compared to pellets striking a tray of ball bearings.

Silicon and germanium are good semiconducting materials. They can be made very pure and both have a very desirable crystalline structure. Electrons can be set free in both of these materials by the action of thermal energy or by striking the materials with photons of light energy.

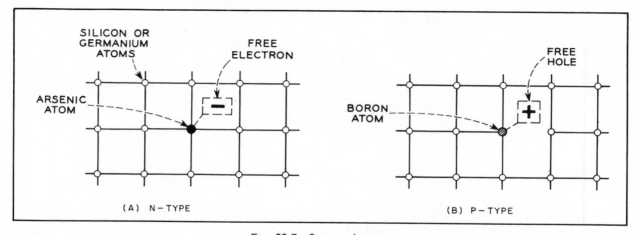

Fig. 23-7   Semiconductors.

Construction of the Solar Battery

Pure silicon is combined with a minute amount of arsenic by a special process in making the solar battery. The initial combination of silicon and arsenic creates an N-type semiconductor. The addition of boron in a later stage of the battery's construction produces a P-type semiconductor. The area of the battery where the two semiconductors meet is referred to as the PN-junction.

1. Ultra pure silicon is "doped" with a minute amount of arsenic. The two materials are melted together and cast in the form of a small ingot, a single crystal of silicon. The addition of the arsenic makes the silicon rich in electrons. Silicon having been treated in this manner is called an N-type semiconductor.

2. The single silicon crystal is then sliced into very small, thin wafers. Rectangular in shape, these wafers are cut about 1/25th of an inch thick by a special crystal cutting machine.

3. The silicon wafers are then polished to a high degree.

4. Boron is added to the surface of the silicon wafer. The wafer is placed in a quartz tube with a vapor containing boron. The boron is diffused into the surface of the silicon. Boron has the effect of reducing the total number of electrons that are potentially free. The boron-treated area of the wafer is said to be a P-type semiconductor. The boron penetrates only a small distance into the wafer, about 1/10,000 of an inch. The area where the N-type semiconductor and the P-type semiconductor meet is called the PN-junction. The PN-junction is the heart of the solar battery.

5. Finally, the wafers are cleaned and then plated so that conducting wires can be attached to the wafer. Connected in series, several wafers are combined to form a solar-energized power source of electrical energy.

steps in the construction of the BELL SOLAR BATTERY

RAW MATERIAL

Arsenic   Purified Silicon

SINGLE CRYSTAL SILICON

Crystal Puller

Silicon Single Crystal Containing Traces of Arsenic

SILICON WAFER

Crystal Cutting Machine

Wafer Sawed From Silicon Single Crystal

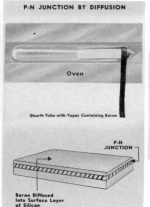

P-N JUNCTION BY DIFFUSION

Oven

Quartz Tube with Vapor Containing Boron

P-N JUNCTION

Boron Diffused Into Surface Layer of Silicon

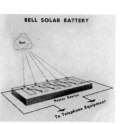

BELL SOLAR BATTERY

Sun

Power Source

To Telephone Equipment

BELL TELEPHONE LABORATORIES

The PN-junction is important to consider. The N-type silicon is by nature electron rich while the P-type silicon is hole rich. When light strikes the solar cell and particularly the PN-junction, photons break electrons away from their atoms. Electrons and holes are created, electron hole pairs. The electric field at the PN-junction forces the electrons into the N-region and the holes into the P-region. The N-region is negative; the P-region is positive. There is a potential difference of about 0.6 volts between the two regions when the battery is exposed to direct sunlight. If a load is connected between the two terminals, the potential will cause electric current to flow. Light en-

ergy has been converted directly into electrical energy by the solar battery.

Fig. 23-8 illustrates how an electric field is built up at the PN junction. (A) The P-region and N-region are separated by an imaginary slide. (B) The slide is pulled out and electrons and holes diffuse across the junction. (C) An electric field quickly builds up at the PN junction to prevent further flow.

Conducting wires are connected to the N-region and the P-region, and a load is placed across the terminals. Electrons flow from the N-region to the P-region.

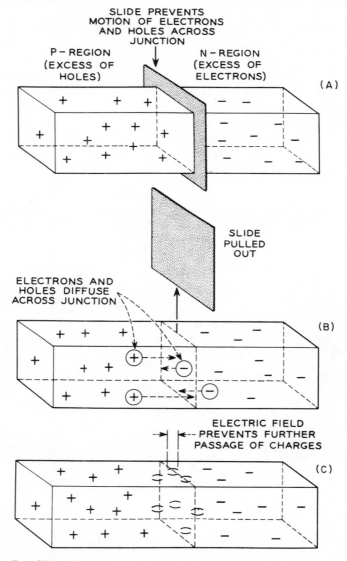

Fig. 23-8   Electric Field Being Built Up at the PN Junction.

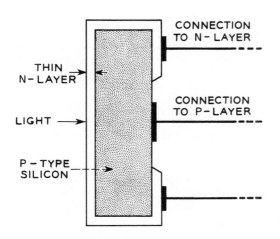

Fig. 23-9   Cross-section Showing Construction
of the Solar Cell.

Fig. 23-10   Solar Batteries Help Provide Electrical
Power for Various Satellite Systems.

Many space applications are developing for solar batteries in addition to the Telstar communications satellite. These batteries are rapidly becoming standard space hardware. Fig. 23-10 shows a large configuration of solar batteries that are grouped with a SNAP device which is also used to produce electricity by direct conversion. SNAP devices (the round ball in the center of the illustration) are discussed later in the unit.

THERMOPHOTOVOLTAIC CONVERTERS

Thermophotovoltaic converters use heat (thermo) to develop an incandescent light source which produces thermal photons (photo) which are directed to a semiconducting material capable of having voltages induced across its terminals (voltaic). This new concept was introduced in 1961 by Pierre Aigrain during his lecture series at the Massachusetts Institute of Technology.

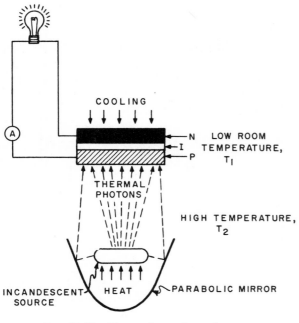

Fig. 23-11   Thermophotovoltaic Converter.

Fig. 23-11 illustrates a thermophotovoltaic (TPV) converter. Remember, in this converter, the photons that excite the semiconducting material do not come from the sun as they do in the case of the solar battery. Rather, they come from a hot incandescent light source. The photons that are generated are directed toward the surface of the semiconducting converter by a parabolic mirror.

Much exploration remains to be done with the TPV process, but it has several advantages. A practical efficiency of 30% to 35% is anticipated. Heat is converted to light through incandescence and, in turn, light is converted to electrical energy by a photocell at room temperature.

FUEL CELL

The fuel cell produces electrical energy directly from chemical energy. To date, these cells are the most efficient non-nuclear devices for generating electrical power. Theoretically, they are 100% efficient but actual gross effi-

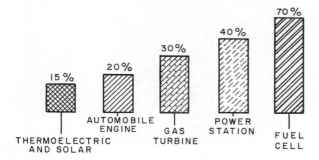

Fig. 23-12 Comparison of Efficiencies of Energy Conversion Systems. Note Superiority of the Fuel Cell.

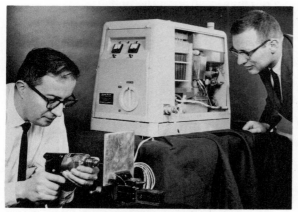

Fig. 23-13 Fuel Cell, Delivering Current for a Power Drill, Operates on Low-cost Liquid Fuel, Methanol.

ciencies of 80% have been recorded. Electrical generation by heat engines can seldom top 40%. The Gemini spacecraft used fuel cells to produce the electrical power needed for operation of the space capsule's electrical/electronic and life-support systems. The nation's moon shot project, Apollo, will also use fuel cells for these purposes.

The applications of fuel cells are not limited to the space program. In 1962, Allis Chalmers announced plans to market a fork lift truck driven by fuel cells. And in 1966, the Esso Research and Engineering Company demonstrated an experimental automated fuel cell battery. If inexpensive materials can be found to replace the expensive platinum catalysts used, such systems may eventually have wide application as a portable power supply.

What, then, is a fuel cell? It is an electrochemical device, not unlike a storage battery which converts chemical energy into electrical energy. But it is different in important ways. Where the storage battery has its fuel built-in as expendable or rechargeable electrodes, the fuel cell has its fuel fed in from outside the cell. And the electrodes of the fuel cell are not consumed as are the battery's.

Fuel Cell Components

In addition to the fuel, the cell contains two electrodes, one positive and one negative, and an electrolyte which serves as the electrochemical connection between the electrodes. Catalysts promote the reaction.

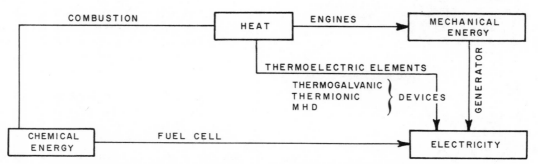

Fig. 23-14 Paths for Conversion of Chemical Energy Into Electrical Energy. Note that Fuel Cells Bypass the Heat Conversion Step.

The electrodes bring the reactants in contact with the electrolyte in a controlled way. They also serve as catalysts and carry the current generated by the reaction. Common electrode materials include silver, nickel, palladium, mercury, platinum and carbon.

The anode is the electrode at which the fuel gives up electrons for delivery to the external circuit.

The cathode is the electrode that gives up electrons to the oxidizer.

The electrolyte provides the means for ionic conduction. Some common electrolytes include potassium hydroxide, sea water, phosphates and zirconates, and others.

No less than two fuels are needed for operation of the fuel cell. One is the oxidant or electron donor, and the other is the reductant, or electron acceptor. A few common oxidants are oxygen, air, chlorine and brine. Reductants include hydrogen, carbon monoxide, natural gas, methanol, zinc, and many others.

Operation

The hydrox fuel cell is one of several types and the one used for space applications. (Others are oxygen-concentration cells, redox cells, hydrocarbon cells, and ion-exchange cells.) The basic principle on which the hydrogen-oxygen (hydrox) fuel cell operates is simple. When hydrogen and oxygen are combined to form water, they release electrical energy.

Fig. 23-16 shows the cell and the basic chemical action. Two hollow porous carbon electrodes are immersed in a potassium hydroxide electrolyte. Hydrogen is pumped into one electrode and oxygen into the other. The hydrogen molecules containing two hydrogen atoms flow into the pores of the negative electrode. Here, a catalyst splits them into two separate hydrogen atoms. The atoms drift to the reaction zone while ions from the potassium hydroxide electrolyte drift toward the same zone. Here the hydroxyl ions (OH) combine to form water (H_2O) and release electrons.

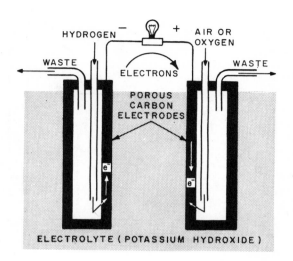

Fig. 23-15   Basic Hydrogen-Oxygen Fuel Cell.

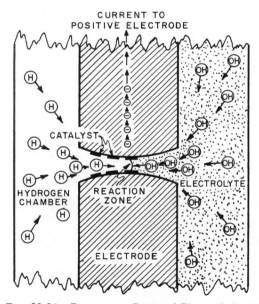

Fig. 23-16   Reaction in Pores of Electrode between Hydrogen and Hydroxide Produces Electrons that Flow Toward Positive Electrode.

The surplus electrons flow toward the positive electrode. Here oxygen atoms have flowed into the pores of the electrode and combined with water molecules from the electrolyte. The incoming electrons combine with this combination to form ions which replace those being used on the negative electrode. Hence, the cycle is continuous.

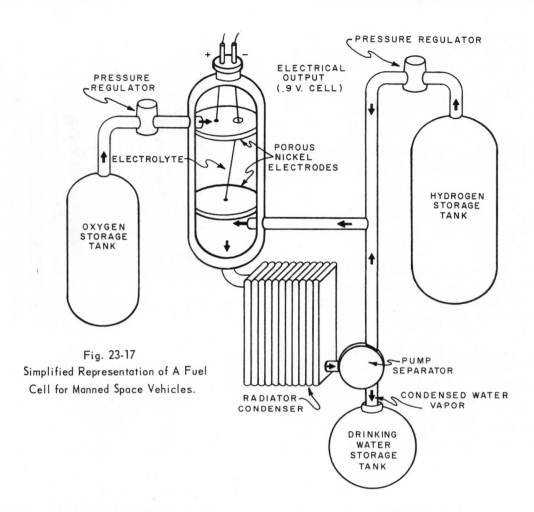

Fig. 23-17
Simplified Representation of A Fuel
Cell for Manned Space Vehicles.

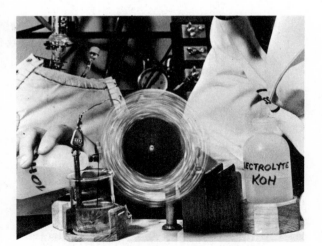

Fig. 23-18   A Demonstration Fuel Cell Simplified for
Student Use. Operates on Methyl Alcohol, Producing
10 Milliamperes current at 0.5 Volts, Enough to
Operate a Transistor Radio.

Advantages and Uses

The advantages of fuel cells are numerous. They have high efficiencies, from 70% to 90%. The reason is that the thermal cycle necessary in the conventional conversion of chemical energy to electrical energy has been eliminated. They have no moving parts, do not produce obnoxious fumes, and they can be built in any size and capacity. They are rugged, require little maintenance, and are efficient in all sizes. An offshoot of the process in the hydrogen-oxygen cell is potable drinking water, which is utilized in such applications as manned space vehicles.

Another important advantage is that when no load is connected to the cell, no energy is consumed.

For spacecraft use, the two fuels, hydrogen and oxygen, are contained in high-pressure, low-temperature tanks which liquefy these gases, conserving valuable space and weight. The gases are put through a heat exchanger which reduces their temperature to about that of the cell before they are introduced to the cell. A fuel cell using methanol and air presents fewer fuel storage problems for land use and has greater promise of approaching a competitive energy cost. Methane is an inexpensive and readily available fuel. Air is free.

The development of inexpensive materials and less complex systems are the goals for further research. And there are many problems which do require such research. One obvious problem is elimination of the water which results from the reaction and weakens the electrolyte. As mentioned earlier, the manned space application will make good use of this water. Another problem is that gases of high purity are required. One attempt to avoid this problem is the Redox cell (reduction and oxidation), Fig. 23-19. Although these and other problems remain to be overcome, the fuel cell holds promise as a significant power source at some future time.

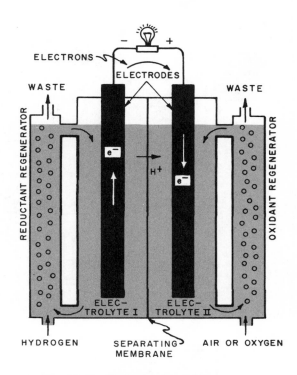

Fig. 23-19   Redox Cell with Membrane Separating Two Electrodes.

THE NUCLEAR BATTERY

The nuclear battery is capable of directly converting the nuclear energy of a strong beta emitter like Strontium 90 directly into electricity. The device is very straightforward, simply a central rod that is coated with the beta-emitting radioisotope -- the source of electrons. The rod is in a vacuum-sealed container and the beta particles, electrons, cross the vacuum gap to a metal sleeve that acts as a collector. The sleeve is connected to the load and current flows.

Space charge effects do not bother the nuclear electrons; they are leaving the radioisotope with a million times more kinetic energy than electrons leaving a thermionic surface.

While nuclear batteries apply voltages of 10,000 to 100,000 volts, the electron flow is only in the microampere range.

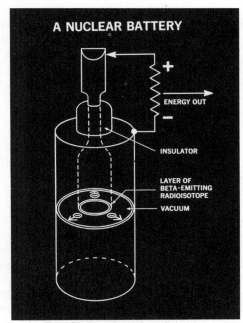

Fig. 23-20   A Nuclear Battery.

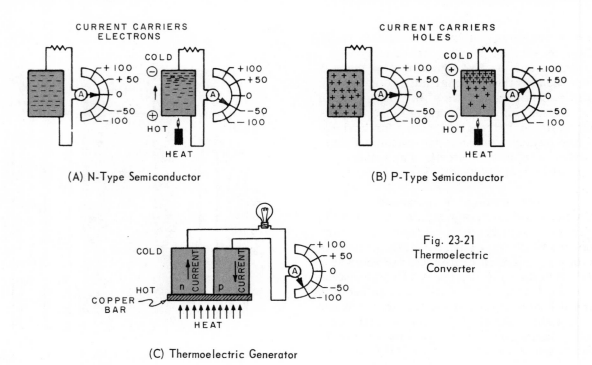

(A) N-Type Semiconductor

(B) P-Type Semiconductor

(C) Thermoelectric Generator

Fig. 23-21
Thermoelectric
Converter

THERMOELECTRICITY

The thermoelectric principle has been known for some 140 years. This principle, one that converts heat directly into electricity, was observed by T. J. Seebeck in 1821. Two dissimilar materials, such as two different wires were joined together and heated. At the unheated terminals a small voltage was recorded. The voltages produced were so small that little practical use could be contrived for the thermoelectric principle until recently when semiconducting materials were developed. Thermocouples made of semiconducting materials are vastly more efficient than earlier thermocouples and they can be put to real use.

A review of semiconductor principles will be helpful in understanding thermoelectricity. The semiconductor does not have many loose electrons that can become current as do metallic materials. The semiconducting material would, in fact, be an insulator if impurity atoms were not introduced into the material. If the impurity atom that is introduced has more than enough electrons to satisfy the valence-bond needs of the surrounding atoms, there is obviously quite an electron surplus. A semiconductor of this type is called an N-type semiconductor, the "N" standing for negative electrons.

If the impurity atoms introduced on the semiconducting material are basically short of electrons themselves and therefore do not have sufficient electrons to satisfy the valence-bond needs of the surrounding atoms, the material is full of positive holes. These "holes" can move within the lattice framework of the material, wandering around much the same as electrons do. But the hole is positive; hence these semiconductors are referred to as "P"-type semiconductors.

Fig. 23-21 shows a simplified version of heat-to-electricity behavior in N-type and P-type semiconductors. When heat is applied to one end of the semiconductor as shown at (A), the electrons in that end increase their velocities and their kinetic energy. These "hot" electrons travel toward the cold end of the semiconductor and pile up. When the circuit is closed, these electrons will flow through the semiconductor as shown, and through the wire from negative to positive.

When a P-type semiconductor is used as in Fig. 23-21 (B), the positive holes are the carriers. The positive holes pile up at the cold end, and electrons flow from the cold toward

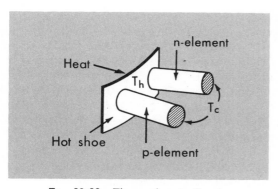

Fig. 23-22 Thermoelectric Couple.

the hot end of the semiconductor, and from negative to positive through the wire as shown.

A thermoelectric generator connects both the types of semiconductors into an effective producer of electrical power. It transforms heat applied from the bottom directly into electrical energy.

The whole operation of producing electricity from a heated thermocouple is still grossly inefficient, in the range of 1% to 5%, whereas a steam power plant might have an efficiency of between 35% to 40%. But still there are distinct advantages to this direct conversion principle. There are no moving parts, so maintenance and lubrication are not a problem. Silence of operation is inherent in the device. Any source of heat can be used to stimulate the flow of electrons and holes,

gas, radioisotopes, and many others. Thermoelectric devices can therefore operate for long periods of time unattended and on this point they find their greatest application; powering navigational buoys, remote weather stations in the polar regions, powering floating weather stations, powering undersea navigational beacons, and so forth.

In the space program, a thermoelectric generator of electricity has several advantages over solar batteries in that it can provide power when the satellite or space capsule is in a shadow and cannot use the sun. Also, when solar cells must pass through radioactive belts they tend to deteriorate. Compared to traditional batteries the thermoelectric device wins in both length of service and weight.

The thermoelectric converters that are used in the space programs of today and those of tomorrow have been largely developed under the SNAP program. This program was initiated by the Atomic Energy Commission in 1956. SNAP stands for Systems for Nuclear Auxiliary Power.

One of the basic concepts of the program was to apply the heat from decaying radioisotopes to the junctions of a thermoelectric mechanism to produce electricity. It might be pointed out that another branch of the SNAP program is at work producing electricity by using small nuclear space reactors as a power source.

Fig. 23-23 A SNAP Generator Being Installed in an Unattended Weather Station 700 Miles from the South Pole.

You may remember that an isotope of an element has the same number of electrons and protons as the element, but it has a different number of neutrons. Therefore, the chemistry of the isotope and the element is the same but their atomic weights are different. Of the many isotopes, some are unstable and decay, emitting three types of radiation — alpha, beta, and gamma rays. These unstable isotopes are called radioisotopes.

Radioisotopes can be used to our benefit in several different ways: (1) They may be directed at a material with the purpose of altering the material, such as destroying cancerous tissue with radiation treatment, (2) They may be used by directing a fixed source at the object to detect the amount of radiation that is reflected or penetrates the object, such as using radioactive material to gage a moving sheet of steel, (3) They may be used in medical and agricultural research where small amounts of radioactive material can be introduced into and followed or traced through a system, and (4) The energy released in radioisotope decay can be used to produce thermal energy. SNAP devices produce thermal energy as the radiation is absorbed.

Scientists have successfully combined the semiconducting thermocouple and the radioisotope. The first of these devices was completed in 1959 — it was the size of a grapefruit and was capable of producing 2.5 watts of electricity for a 90-day period. It used polonium-210 as the radioisotope fuel. Other SNAP devices of various power outputs and longevity have followed:

DESIGNATION	USE	POWER (watts)	GENERATOR LIFE	ISOTOPE
SNAP-3	Demonstration Device	2.5	90 days	Polonium-210
SNAP-3A	Satellite Power	2.7	5 years	Plutonium-238
	Axel Heilberg Weather Station	5	2 years min.	Strontium-90
SNAP-7A	Navigational Buoy	10	2 years min.	Strontium-90
SNAP-7B	Fixed Navigational Light	60	2 years min.	Strontium-90
SNAP-7C	Weather Station	10	2 years min.	Strontium-90
SNAP-7D	Floating Weather Station	60	2 years min.	Strontium-90
SNAP-7E	Ocean-bottom Beacon	7.5	2 years min.	Strontium-90
SNAP-7F	Offshore Oil Rig	60	2 years min.	Strontium-90
SNAP-9A	Satellite Power	25	5 years	Plutonium-238
SNAP-11	Moon Probe	21-25	90 days	Curium-242
SNAP-13	Demonstration Device	12	90 days	Curium-242
SNAP-15A	Military Use	0.001	5 years	Plutonium-238
SNAP-17	Communication Satellite	25	5 years	Strontium-90
SNAP-19B	Nimbus-B Weather Satellite	30	5 years	Plutonium-238
SNAP-21	Deep Sea Use	10	5 years	Strontium-90
SNAP-23	Terrestrial Uses	60	5 years	Strontium-90
SNAP-27	Lunar Landings	60	5 years	Plutonium-238
SNAP-29	Various Missions	500	90 days	Polonium-210

The SNAP device is made up of (1) an outer shell or can that protects the internal components. (2) A radioactive shield under the shell. If alpha-emitting fuels are used there is little need for the shield, but if gamma radiation is present a massive shield may be needed. (3) Thermoelectric converters or thermocouples that convert a portion of the decay into electricity, and (4) Fuel capsule of radioactive material located at the very center of the generator.

The radiation particles themselves do not escape from the device; in fact, most all are trapped inside the fuel capsule itself where their vast kinetic energy is converted into heat energy. Of the heat energy that is produced, only 5% to 10% is used to produce electricity, the rest eventually escapes to the outer shell and to the media the device is operating in.

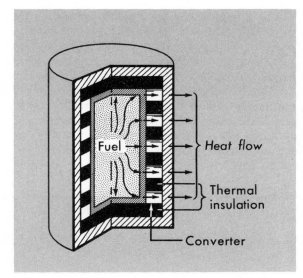

Fig. 23-24   Radioisotope Generator Showing Unavoidable Heat Loss.

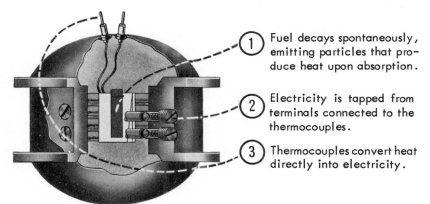

① Fuel decays spontaneously, emitting particles that produce heat upon absorption.

② Electricity is tapped from terminals connected to the thermocouples.

③ Thermocouples convert heat directly into electricity.

Fig. 23-25   Operation of Thermoelectric Generator Powered by Radio Isotopes.

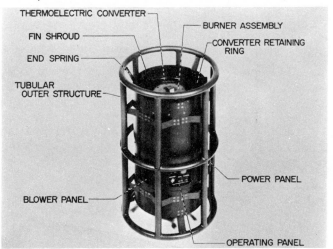

Fig. 23-26   A Thermoelectric Device that Can Operate on Leaded Gasoline, Jet Fuel or Diesel Fuel.

Thermoelectric devices that employ more conventional heat sources have also recently come out of the laboratory and into production. These units employ propane or natural gas fuels fired through an enclosed burner. The primary advantage of hydrocarbon fuels versus radioisotope fuels is that of cost. Radioisotope fuels are fantastically expensive. Hydrocarbon fuels can even compete very nicely with conventional primary batteries. Remote site power can be provided for $2 to $3/kw.-hr. using a gas-fired thermoelectric generator while primary batteries may cost $10 to $18/kw.-hr.

Like the SNAP devices, the gas-fired thermoelectric devices are excellent for remote, unattended service. Of course, periodic replenishment of the fuel is necessary. The Minnesota Mining and Manufacturing Co. is currently producing commercial units that range in output from less than one watt to 50 watts, and larger wattage units are on the way, Fig. 23-26.

THERMIONIC CONVERTERS

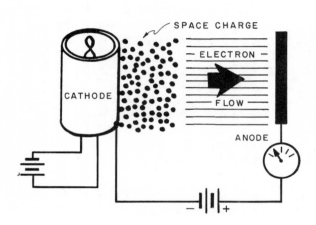

Fig. 23-27   Space Charge Effect.

Thermionic conversion, or the generation of electricity from heat, is based on Thomas Edison's inadvertent discovery that electrons are emitted from a heated cathode, the Edison Effect. However, such boiling off of electrons from the hot cathode to the cold anode (collector) results in an electrostatic charge of negative electrons forming near the cathode. This "space charge" repels other electrons and blocks their flow. Depending on the methods used to overcome the space charge barrier, three main types of thermionic converters have been developed:

1. The vacuum close-spaced diode

2. The magnetic triode

3. The cesium plasma diode

Vacuum Close-Spaced Diode

This vacuum tube places the cathode and anode very close together to avoid the detrimental effect of the space charge. Close-spaced diodes were first developed at MIT in 1956 and were called "thermoelectron machines." They are heat engines which use the electron gas to deliver useful work as a result of passing through the electrostatic field.

The efficiency of these devices is from 12% to 14% and they weigh the least of all energy converters. However, the cathodes must operate at high temperatures which shorten their life considerably and reduce efficiency.

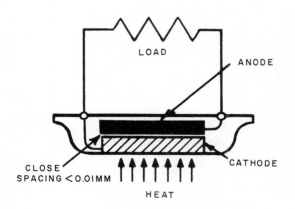

Fig. 23-28   Vacuum Close-Spaced Diode.

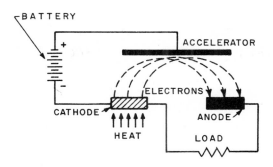

Fig. 23-29 Magnetic Triode.

Magnetic Triode

This device reduces the space charge effect by using crossed electric and magnetic fields. A hot cathode and a cold anode are placed in the same plane and close together. Above them, an accelerating anode is placed. A battery produces an electric field between the cathode and anode. An external magnetic field is superimposed by placing the entire device between the poles of a large magnet.

The electron emitted from the hot cathode speeds toward the accelerator. But the magnetic field deflects it toward the surface of the anode.

During its acceleration, the electron gains kinetic energy from the electric field. When it falls back to the anode, it returns that energy to the field. The result is that the energies of the electrons arriving at the anode are the same as they were when they left the cathode. In theory then, the magnetic triode should produce more power than the close-spaced diode. However, due to a number of electron losses, such is not the case.

In general, thermionic converters are low-voltage, high-current generators. Their main applications to date have been in space vehicles.

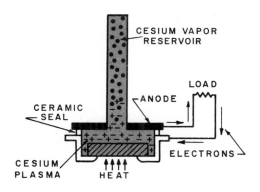

Fig. 23-30 Cesium Plasma Diode.

Cesium Plasma Diode

Use of an ionized gas is another way of counteracting the space charge. In diodes filled with cesium plasma, the positive ions neutralize the negative space charge and allow the electrons to flow toward the anode. A single ion can effectively cancel several hundred electrons.

As in the vacuum close-spaced diode, the cathode must operate at exceedingly high temperatures which shortens its life. Other fabrication difficulties are involved with the corrosiveness of the cesium vapor. Efficiencies from 10 to 25% have been achieved.

Intensive thermionic research is underway with the goal being the development of reliable converters for space use. These thermionic devices might be heated by the sun, by decaying radioisotopes or by the heat of a fission reactor. If scientists could simply wrap the thermionic converter emitter around a reactor fuel element, the result would be an ideal combination. However, many problems are involved in developing materials that can withstand the extremely high operating temperature (3092°F) in the physical construction of the device, and in the nuclear physics factors that must be considered. Much work remains to be done in the area of thermionic conversion devices before practical applications can be made.

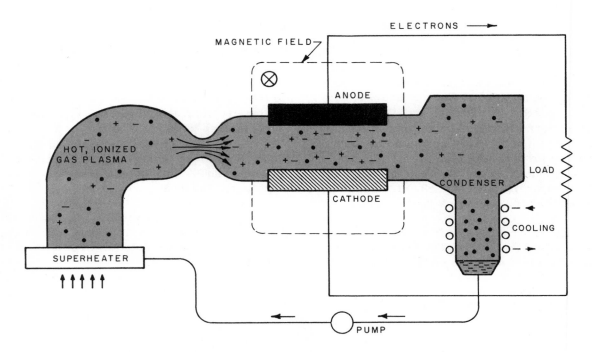

Fig. 23-31   Magnetohydrodynamic (MHD) Generator.

MAGNETOHYDRODYNAMIC GENERATOR

The magnetohydrodynamic power generator makes application of the basic principles of electrical generation that were discovered by Faraday in the early nineteenth century — simply, that by moving an electrical conductor through a magnetic field or vice versa, a voltage will be induced. Steam, water, and gas turbines are often used in this way, applying their energy to rotate a conductor (the armature) through a magnetic field. In the MHD (magnetohydrodynamic) generator, a pressure-driven conductive gas replaces the conductive copper wire of the armature that moves through the magnetic field of the normal generator.

The idea of a gaseous conductor sounds simple at first, but in order for a gas to be an effective conductor of electricity it must be ionized and this can be accomplished best only by intense heat. In the ionization process, electrons can be boiled off and become free electrons, thus leaving the gas atom with a positive electrical charge; we say it is "ionized." These free electrons can drift in the gaseous conductor, some recombining with ionized atoms, but many are free at any given time. Many free electrons are drawn to and picked up by the anode and delivered to the load.

The hot gaseous ionized fluid is referred to as "plasma." But simply burning a chemical fuel such as powdered coal, oil, or natural gas with ambient air will not produce temperatures that are sufficient to promote the slight ionization that is necessary. To solve this problem the gas is "seeded" with a material that will ionize easily, such as potassium. When seeded with potassium, ionization will result at ordinary oxygen combustion temperatures of 2500°-3000°K. To insure efficiency and to obtain the high temperatures needed, the air is often oxygen-enriched and preheated.

Temperatures of the magnitude required in an MHD generator would be destructive to ordinary turbine generators but in the MHD generator there are no moving parts that the gaseous stream is directed toward. In fact, in the actual production of electricity there are no parts at all.

The combustor itself is very similar to a rocket engine, Fig. 23-32. The hot plasma exits from the nozzle into a duct of increasing diameter. A magnetic field surrounds this duct and the motion of a conductive plasma gas through this field induces voltage across the electrodes. In actual practice the electrodes are segmented, each segment being led to the load, in order to minimize the Hall effect or the tendency of the conductor to set up a magnetic field around itself, which would serve to cancel out the force of the field coil.

The MHD generator is ideally suited for producing large quantities of electricity at central power stations. When these units are installed they will incorporate several other principles and machines to attain efficiencies of an expected 50%, ten points higher than conventional plants are now capable of. To attain high efficiency, hot gases cannot simply be exhausted. If the hot plasma were exhausted away it would represent a tremendous loss. A portion of the heat can be used in a regenerator to preheat the incoming air. Another portion of the hot gases can be used in a steam cycle whose turbine turns both a compressor and a conventional generator. Even the "seed" material could be reclaimed from the ash of the exhaust gases. This type of operation is referred to as an "open" cycle since the plasma gas is eventually exhausted from the mechanisms when as much energy as is practical is derived from it.

The possible use of nuclear reactor power for MHD generators presents another area of promise. However, the temperatures attained are in the range of 100° to 1250°K and insufficient to cause thermal ionization.

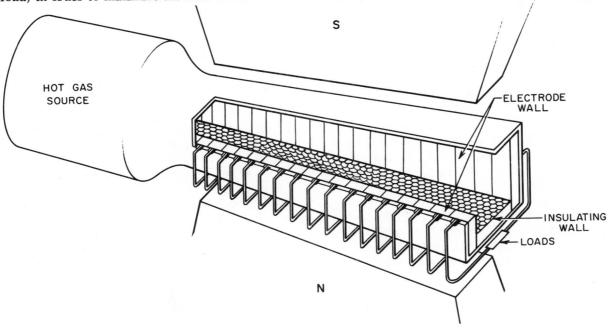

Fig. 23-32 Schematic Drawing of MHD Generator (Normal Segmented Hookup)

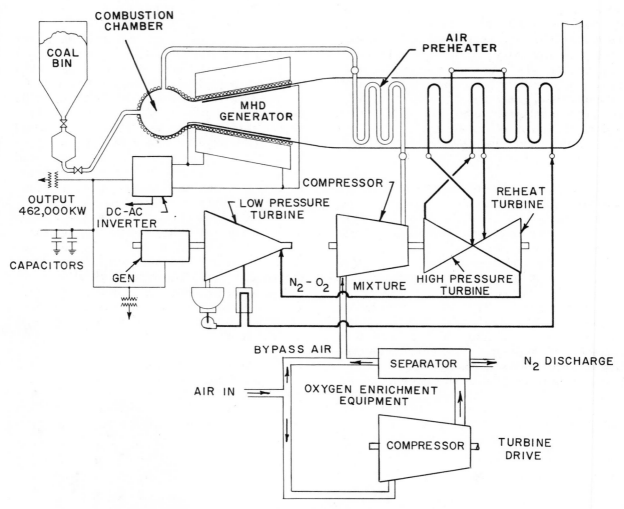

Fig. 23-33 Design for a Coal-Burning, Open-Cycle MHD Power Plant.

Either of two routes can be taken to solve this limitation: (1) reduce the temperatures at which an MHD generator will operate or (2) build nuclear reactors that will produce greater heat intensities.

Current findings indicate that future nuclear-powered MHD generators will be of a closed cycle design. The plasma, probably argon or helium, will be heated by the reactor but not interact with it. The ionized gas will pass through the duct but will not be exhausted from the system. Rather, it will be drawn off at the discharge end of the duct, reheated and expanded through the nozzle again. Safety factors of radioactive contamination present problems for the design of these generators.

At present, the prime area of development of MHD generators is directed toward the generation of electrical power at the central power station. Large MHD generators can produce huge amounts of power at high efficiency. The intense heat required to ionize gases always presents the problem of developing materials and components that can take the heat for long periods of time. Concentrated work on MHD generators has been exerted for only a few years. Much progress has been made — the future holds a great deal of promise.

GENERAL STUDY QUESTIONS

1. What common direct energy conversion device has served us for many years?
2. What energy form is produced by most all direct energy converters?
3. What are the general requirements for space exploration electrical/electronic power supplies?

SOLAR CELL

1. Explain how a power source such as coal can be traced to the sun.
2. Name three early devices for converting the sun's energy directly into electricity.
3. What group is responsible for the latest improvements in the solar battery?
4. What is the efficiency of present day solar batteries?
5. What applications have been found for the solar battery?
6. What is a semiconductor?
7. How can a semiconductor be made conductive?
8. What type of semiconducting material is used in the solar battery?
9. What is the arsenic "doped" region called?
10. What is the boron "doped" region called?
11. What is the PN-junction?

THERMOPHOTOVOLTAIC CONVERTER

1. What is the source of the photons that excite the semiconducting material of the thermophotovoltaic converter?
2. Where and when was this converter introduced?

FUEL CELL

1. How does the efficiency of the fuel cell compare with all other non-nuclear energy conversion systems?
2. Does a fuel cell store the energy of a fuel or does it furnish power as it consumes a fuel?
3. What potential uses are seen for the fuel cell?
4. Has the fuel cell seen any practical application to date? Explain.

NUCLEAR BATTERY

1. Explain the basic operating principle of the nuclear battery.
2. Why are space charge effects of little consequence to a nuclear battery?

THERMOELECTRICITY

1. Explain the basic principle of a thermocouple.
2. Why are semiconducting materials used in modern thermoelectric devices?

3. What characteristics of a thermoelectric device makes it suitable for use at remote, unattended locations?

4. How can radioisotope materials be used for thermoelectric devices?

5. What is the SNAP program?

THERMIONIC CONVERTER

1. Explain the term "space charge." What has it to do with the problems of thermionic converters?

2. Describe the basic principle upon which the vacuum close-spaced diode operates.

3. What is the major disadvantage of the vacuum close-spaced diode?

4. How does the magnetic triode solve the problem of space charge?

5. On what principle does the cesium plasma diode operate?

MAGNETOHYDRODYNAMIC GENERATOR

1. Explain the principle on which the MHD generator operates.

2. What is "plasma"?

3. What is the difference between the open cycle principle and the closed cycle principle?

4. What difficulties are encountered in MHD generator design?

CLASS DISCUSSION TOPICS

● Discuss the energy conversion matrix and trace the energy paths for several prime movers and their applications.

● Discuss the inherent advantages of energy conversion devices that contain no moving parts.

● Discuss the types of educational backgrounds that are needed in the development of space age power.

● Discuss the sun as the ultimate source of power.

● Discuss the potential uses of the solar battery, particularly in underdeveloped parts of the world.

● Discuss the use of the fuel cell in manned space vehicles and how the byproducts of the reaction can be used.

● Discuss the advantages of thermoelectric devices for use in unattended sites where power is required.

● Discuss thermoelectric devices in addition to those produced under the SNAP program.

● Discuss the term "boiling off" electrons.

● Discuss some of the problems that face engineers in the thermionics field.

● Discuss how an MHD generator is similar to a conventional generator.

Unit 24
OCCUPATIONAL OPPORTUNITIES IN POWER

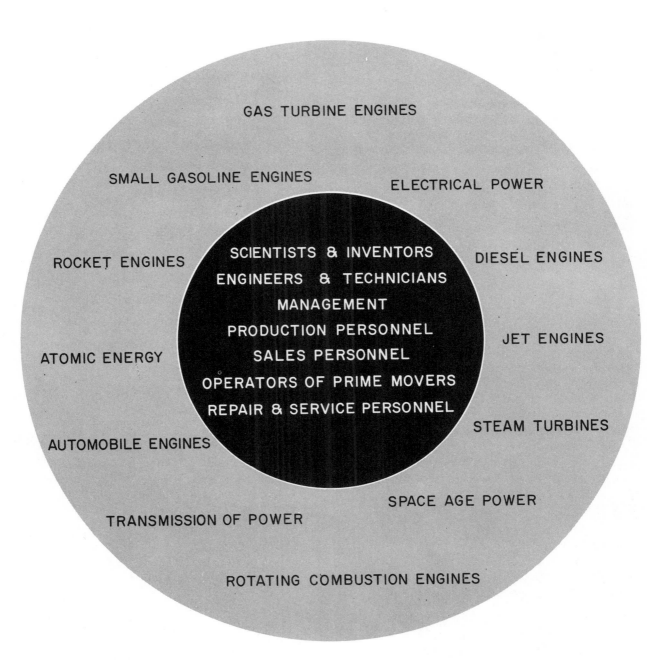

Fig. 24-1   A Wide Range of Skills, Talents, and Interests Lie Within the Field
of Power Technology.

The scope of "Power Technology" is so broad that it touches the lives of everyone. Every person is a consumer of power; it is just an inescapable part of our way of life. The automobile, electrical household appliances and machines, power lawnmowers, and recreational equipment are all woven very thoroughly into the fabric that we call "The American Way of Life." So everyone has a "Future in Power," as a consumer if not as a vocation or life's work.

Sound understandings of the world of power technology will certainly help to make a person a more intelligent consumer. And, further, these understandings will increase one's appreciation and awareness of all that goes on around him. Concepts of power and its relationship to the community and its life surely help a person to enjoy life to a greater degree.

But what about occupations in power technology? Is there a future for large numbers of persons? What kind of skills, aptitudes, and talents are needed? These questions might first lead us back to the definition of power technology — energy sources and the machines that convert energy into useful work. The definition, in turn, leads to the several units that lie within the covers of this book:

- Small Gasoline Engines
- Automobile Engines
- Diesel Engines
- Jet Engines
- Rocket Engines
- Gas Turbine Engines
- Rotating Combustion Engines
- Electrical Power
- Mechanical Transmission of Power
- Fluid Transmission of Power
- Steam Turbines
- Atomic Energy
- Space Age Power

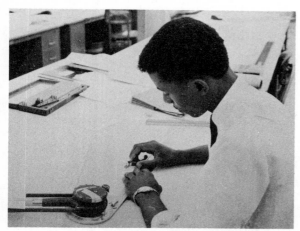

Fig. 24-2   Engineer at Work at Aero Jet General Corp.

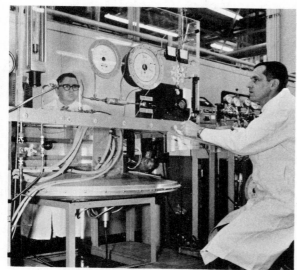

Fig. 24-3   Radio Frequency Plasma Research Done by Scientists.

Fig. 24-4   Work Progresses in Spacecraft Assembly Within a Class 6 Clean Room at McDonnell Douglas Corp.

The vastness of the field and the vastness of opportunities stagger the imagination. Millions upon millions of trained men and women are needed to meet the needs of our power-consuming society.

To the uninitiated, a career in power technology could have as narrow a meaning as a career as an auto technician, and it might be added that this is one phase of power technology: there are over 700,000 auto technicians. However, in its broad sense, the field of power technology would also encompass the electrical engineer and the atomic physicist.

If a person will but look around his own community he can see how power employs many persons, some in a very direct way such as

service station managers or the employees of the public utility company. Others are employed in power in a more indirect way — perhaps operating trucks, trains, buses, or aircraft in the community's transportation network.

Every spoke of the power technology wheel requires persons with special talents, interests, and skills.

Scientists, researchers, and inventors who conceive ideas and principles.

Engineers and technicians who develop scientific principles into practical applications.

Management which is skilled in organizing large-scale endeavors in power.

Production personnel who can manufacture the ever-lengthening list of prime movers.

Sales personnel who understand power and who can fill the power needs of business, industry, and individual consumers.

Operators of prime movers which would include the professional airplane pilot, the railroad engineer, the truck driver, the stationary engineer of the power plant, and others.

Repair and service personnel who have the skill needed to keep prime movers operating efficiently and safely.

Fig. 24-5   A High Degree of Skill is Necessary in the Final Assembly of Aircraft.

Fig. 24-6   A Highly Skilled Employee Operates this Numerically Controlled Milling Machine at McDonnell Douglas Corp.

Fig. 24-7   Workmen Assembling a Large Gas Turbine Engine.

The entire list of industries producing fuel for prime movers might also be listed since they represent the "sources of energy" consumed by prime movers: coal, natural gas, petroleum, nuclear fuel, and exotic fuels for space age power.

Where any individual might fit into the world of power depends on his interest and ability. Would you like to sell automobiles, or would you rather have a wrench in your hand and repair automobiles, or would you rather operate an auto parts business? Would you prefer to work as a diesel mechanic or as a railroad engineer? Does the idea of being a space engineer appeal to you, or would you rather be a member of the team that actually manufactures the many component parts of a space vehicle? Would you like to work as an engineer, a manager, a production worker, or a skilled worker in a manufacturing plant that produces a type of prime mover?

It is not the purpose of this unit to discuss in detail the thousands of occupations directly related or allied with power technology but it can certainly stimulate interest and a quest for further information. Many occupational books and pamphlets deal directly with the more common occupations that lie in the field and these are on file in school and public libraries.

Another way of viewing your future in power technology is to look at the training that is required for various job levels. This is one method that is commonly used for classifying occupations.

Professional

College education required; four or more years of formal education beyond the high school level. Scientists, engineers, researchers, high level managers, and executives are among these.

Technical

Some college or technical training required, but not necessarily a college degree. Draftsmen, engineering assistants, experimental mechanics, and technicians are among these.

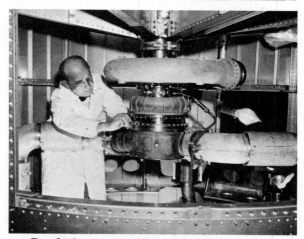

Fig. 24-8   Aero Jet Personnel at Work on Titan II Rocket Engine Components.

Skilled Workers

Lengthy preparation through on-the-job training or through apprenticeship programs is required. Aircraft mechanics, diesel mechanics, auto mechanics, machinists, and tool and die makers are among these occupations.

Semiskilled Workers

Considerable on-the-job training is required. Generally, these persons are machine operators and are employed in every major industry.

Unskilled Workers

Little on-the-job training is required. Helpers, laborers, and gas station attendants fall into this occupational category.

In terms of earnings, the occupations which require the most training, the most creative effort, the most attention to detail, and those that carry the greatest responsibility provide the greatest financial rewards. The financial aspect of an occupation is important but a person must ask himself very honestly: is the occupation suitable for my mental and physical ability and will it provide me with a lifetime of satisfying work? Opportunities abound — consider these additional facts:

1. One person in seven earns a living from the manufacture, sale, service, and use of motor vehicles.

2. One business out of every six is automotive in nature.

3. More than 2,000,000 persons earn a living operating devices driven by prime movers: trucks, buses, trains, taxis, etc.

4. Industrial production workers have at their disposal an average of 1249 hp. per worker in installed electrical motors and prime movers.

5. More than 1,400,000 persons are employed in aircraft, missile, and spacecraft manufacturing.

6. The rapidly developing field of atomic energy already employs more than 200,000 persons.

GENERAL STUDY QUESTIONS

1. Explain how a salesperson might be involved in power technology.

2. Prepare a list of several occupational groups that are needed to bring a prime mover into the economy.

3. Briefly explain the educational requirements of the following job classifications:

    √ Professional

    √ Technical

    √ Skilled Workers

    √ Semiskilled Workers

    √ Unskilled Workers

CLASS DISCUSSION TOPICS

● Discuss how power technology can touch the lives of every person.

● Discuss occupations in your community or neighborhood that lie in the broad field of power technology.

● Discuss several factors that should be considered in selecting an occupation.

● Discuss how technology has upgraded the skills needed by today's and tomorrow's work force.

UNIT 25

POWER TECHNOLOGY AND THE ENVIRONMENT

Power has had much to do with the present high degree of civilization, but on the other hand, we are depleting natural resources on an ever-increasing scale in order to utilize energy-converting machines and other devices of power technology. The phenomenal increase in the use of power has created a good deal of the present environmental and ecological problem. Consider that in 1945 there were 31 million automobiles in the United States, and in 1970 there were 104 million autos which burned up to 4 times as much gasoline as those produced in 1945. In the consumption of electric power the record has been similar. In 1945 we consumed 271 billion kilowatt hours of electricity, and as recently as 1969 we used electricity at the rate of 1552 billion kilowatt hours per year. Only in the past few years have we realized that we are paying a huge price for all this power: a price in mineral depletion, defacement of the landscape, and pollution of air and water.

The scope of the problem is vast, and its many parts are interrelated. At the beginning of our discussion, it is well to define three important terms used.

Environment: Our environment is the sum total of the things, conditions, and influences that affect the development of our lives.

Pollution: Pollution is waste that makes the environment unclean or foul. Pollution can harm our health, safety, and even technology itself. Pollution also offends our moral and aesthetic sensibilities, decreasing the quality of our lives.

Ecology: Ecology is a branch of biology that concerns itself with the relationship of organisms and their environment. Ecology is the study of the interdependence of all forms of life.

THE GOVERNMENT AND POLLUTION

A few steadfast conservationists have tried through the years to inform the citizens of a rapidly approaching environmental crisis. They experienced moderate success in arousing the public and in securing the passage of

Fig. 25-1A Man Has Only One World —
The Resources are Limited.

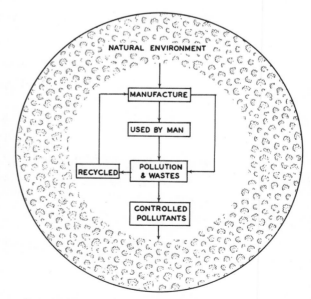

Fig. 25-1B Our Natural Environment can be Saved
With a System that Recycles Resources
and Controls Pollutants.

legislation aimed at environmental problems. The late Rachel Carson's book, Silent Spring, published in 1962, did much to stimulate widespread interest in the environment. Later in the decade, oil spills, which fouled beaches and destroyed marine life, and smog, which endangered the health of people living in urban areas, awakened the public to the fact that the environment was indeed being threatened.

Actually, the federal government expressed concern with the passage of the Air Pollution Control Act in 1955 which funded air pollution research. The Clean Air Act of 1963 poured millions more into pollution problems and shifted the emphasis to developing, establishing, and improving pollution prevention and control programs.

In 1965 the Motor Vehicle Air Pollution Control Act provided for the establishment of emission standards for new motor vehicles. In 1968 the auto industry began to meet federal emission standards, which have become increasingly stringent in recent years.

Federal involvement in all forms of pollution continued to grow, and in 1970 a master agency called the Environmental Protection Agency (EPA) was formed. The EPA has authority over municipalities, states, and industries in pollution control of air, water, radiation hazards, solid wastes, and pesticides.

The Council on Environmental Quality (CEQ) and the National Oceanic and Atmospheric Agency (NOAA) were established in 1970. The CEQ is assigned the task of supervising the government itself. This agency checks out all projects, plans, and proposals that may affect the environment. The NOAA is designed to be a superior scientific agency. It determines just how badly air and water are already polluted and suggests what can be done about it. In addition, the NOAA determines existing resources in the sea and air and explores ways in which man may eventually modify energy to shape the weather.

AIR POLLUTION

The major sources of air pollution are (1) transportation, including the automobile, (2) both industrial and residential heating plants, (3) refuse burning, (4) pollution from industrial processes, and (5) utilities, especially coal-burning electrical power generation stations. These sources of pollution are highly concentrated in urban areas where 70 percent of the population is also concentrated.

People are quick to single out the billowing industrial stacks as the prime polluter, but the facts do not allow the individual to escape the effects of his own pollution: his furnace, incinerator, power mower, and the automobile, which is the largest single source of air pollution.

Several types of air pollutants have been identified. Carbon monoxide (CO) is the colorless, odorless, and tasteless killer in auto exhaust. Other pollutants include sulfur oxides, carbon dioxide (CO_2), lead, nitrogen oxides, assorted hydrocarbons, ash, mists of oils, aerosols, metallic oxides, and various soots. At this point researchers are just beginning their list of harmful air pollutants.

Property damage caused by pollution has a price tag of ten billion dollars per year. Photochemical smog is partially produced by the reactions of nitrogen oxides with oxygen in the sunlight. It is agreed that smog produces damage in the human body, but specific research results are lacking. It can be said, however, that pollutants do aggravate all types of respiratory conditions. Millions of people have experienced the effects of smog: stinging eyes, irritated nasal passages, and breathing difficulties. Breathing polluted air over a lifetime may well be costing us far more than we now realize.

SOLUTIONS TO AIR POLLUTION

Air pollution can be markedly reduced. The big stacks of industry can be equipped with electrostatic precipitators, centrifugal collectors, filters, baffles, and water scrubbers to reduce pollution hazards. These devices are expensive, but industries are moving rapidly to control the emissions from their stacks. Nuclear-powered electrical plants are the cleanest power source, but even nuclear reactors pose environmental threats from radiation and thermal pollution.

The individual himself can reduce air pollution. The cumulative power demands of the public are reflected at the power station as it consumes mountainous stockpiles of coal. Instead of loading the home with labor-saving devices to the point where it takes on the appearance of a miniature factory, the family can do more under their own power. Eliminating the inefficient backyard incinerator can be done with little inconvenience to anyone. Burning a cleaner fuel for residential heating also reduces pollution. Natural gas is quite clean, but it is also in short supply. The most plentiful fuel is coal, but the cheapest coal is high in sulfur content. Taking the sulfur out of coal before burning it is an expensive process. Much of the fuel oil that is priced competitively with coal is also high in sulfur content. Home heating oil has had most of the sulfur removed, but it costs about twice as much as the heavier oils.

SOLUTIONS TO AUTOMOBILE POLLUTION

Many environmentalists favor the approach of de-emphasizing the auto and its place in today's lifestyle. In other words, they advocate enforced return to mass transit systems — possibly even to the point of prohibiting the auto from the most congested areas of a city. The millions of autos in a metropolis pollute the air, take acres of parking space, require fantastic highway complexes, and are generally inefficient in moving people. Ranked in order of efficiency of people-moving, the bus, train, bicycle, and walking are all ahead of the automobile.

No doubt we will experience a renewed interest in mass transit, but there is little evidence that we, as a people, are about to willingly give up the family car. It is so woven into our economy and our lives that life without the car is as unpalatable as life with polluted air. Hence, the automobile must be cleaned up. Spurred on by the pollution reduction standards of the Environmental Protection Agency, automobile manufacturers are doing just that.

Automobile pollution comes from three principle sources: (1) vapors from the crankcase, (2) fuel vaporization from the gas tank and carburetor, and (3) combustion byproducts from the tailpipe.

Crankcases were vented to the atmosphere until it was discovered that the emissions from crankcases are pollutants, about 20 percent of the total problem. These pollutants are unburned hydrocarbons from blowby along the cylinder walls which pass into the crankcase. Since 1963 the crankcase pollution problem has been brought under effective control with positive crankcase ventilation systems. Positive crankcase ventilation recirculates crankcase gases through the engine's burning cycle, so that they do not escape into the atmosphere.

About 20 percent of the hydrocarbon pollution has been traced to the carburetor and gas tank. Gasoline is an ideal fuel because it vaporizes easily; unfortunately, pollutants are created when the gas evaporates from the gas tank and carburetor. Tanks and carburetors must be vented to prevent the buildup of pressures. Prior to pollution controls, they were vented to the atmosphere. Everyone has seen these vapors shimmering in the sun as the gas station attendant fills a gas tank. Most vaporization occurs just after the engine has stopped, and the hot engine evaporates the gasoline left in the carburetor. This kind of evaporation is now controlled, figure 25-2. In one system vapors are collected and stored in the crankcase to be used when the engine is restarted. In another system the vapors are collected in a cannister of activated charcoal granules for later use.

EVAPORATIVE CONTROL SYSTEMS

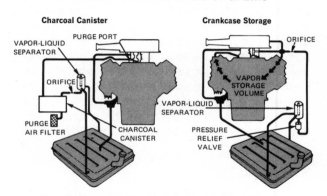

Fig. 25-2 Evaporative Control Systems Eliminate Pollution from the Carburetor and the Gas Tank.

ENGINE MODIFICATION

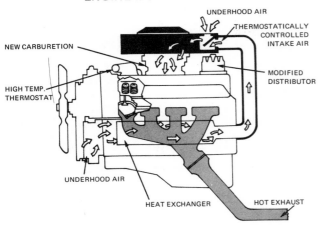

Fig. 25-3 Modification of the Engine's Fuel Induction and Cooling System Reduces Hydrocarbon Emissions From the Tailpipe.

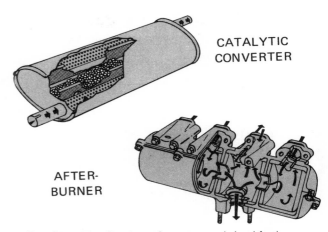

Fig. 25-4 The Catalytic Converter and the Afterburner are Methods of Further Reducing Tailpipe Emissions.

Exhaust emissions contain hydrocarbons, oxides of nitrogen, carbon monoxide, lead particles, and other pollutants. The more completely the fuel is burned, the less the pollution. Controls for exhaust emissions are already in effect, but they must continue to improve to meet standards of the future. Manufacturers are designing combustion chambers, fuel induction systems, ignition systems, and other parts that improve the combustion process, figure 25-3.

Afterburners may be used to insure more complete combustion. In an afterburner, figure 25-4, air is pumped into the exhaust manifold to burn the exhaust more completely before it is released to the atmosphere.

Many manufacturers are approaching further emission control with the catalytic converter, figure 25-4. The catalytic converter chemically changes the hydrocarbons and carbon monoxide emissions into water vapor and carbon dioxide. This device depends on the use of lead-free gasoline. Tetraethyl lead is now blended into gasoline to improve its octane rating. Lead does not hurt the engine. In fact, metallic lead acts as a lubricant for valves, and while lead is harmful as a pollutant, experts have not established the extent of lead hazard in emissions. It has been found, however, that lead is very destructive to catalytic converters. It is necessary to remove the lead from gasoline or to greatly reduce the amount present.

If manufacturers fail to reduce pollution by redesign of engine parts, another option is to replace the conventional auto engine with an engine that pollutes less. Such engines include diesels and turbines. Electric autos pollute hardly at all. To date none of the alternatives provide the operating characteristics of the conventional automobile engine. However, it is possible that a new engine may equal the performance of the auto engine and be cleaner. If and when this happens, the internal combustion engine may be phased out as the propulsion unit for the auto.

GENERAL STUDY QUESTIONS

1. Explain how the use of power has contributed to pollution.

2. Define the following terms:

    a. environment   b. pollution   c. ecology

3. What is the function of the Environmental Protection Agency?

4. Name five major sources of air pollution.

5. List five air pollutants.

6. Explain why rapid transit systems pollute less than automobiles.

7. What are the three sources of pollution from the automobile?

8. List three features of auto design that reduce pollution.

CLASS DISCUSSION TOPICS

● Discuss the environmental problems of your community and geographic region.

● Discuss what personal sacrifices you will have to make in order to solve local pollution problems.

● Discuss the costs that may be involved in correcting environmental problems in your community.

● Discuss the advantages and disadvantages of several automobile power plants other than the conventional automobile engine.

● Discuss current legislation on pollution and the environment.

APPENDIX

TROUBLESHOOTING CHART

ENGINE FAILS TO START OR STARTS WITH DIFFICULTY

Cause	Remedy
No fuel in tank	Fill tank with clean, fresh fuel.
Shutoff valve closed	Open valve.
Obstructed fuel line	Clean fuel screen and line. If necessary, remove and clean carburetor.
Tank cap vent obstructed	Open vent in fuel tank cap.
Water in fuel	Drain tank. Clean carburetor and fuel lines. Dry spark plug points. Fill tank with clean, fresh fuel.
Engine over-choked	Close fuel shutoff and pull the starter until engine starts. Reopen fuel shutoff for normal fuel flow.
Improper carburetor adjustment	Adjust carburetor.
Loose or defective magneto wiring	Check magneto wiring for shorts or grounds; repair if necessary.
Faulty magneto	Check timing, point gap, and if necessary, overhaul magneto.
Spark plug fouled	Clean and regap spark plug.
Spark plug porcelain cracked	Replace spark plug.
Poor compression	Overhaul engine.

ENGINE KNOCKS

Cause	Remedy
Carbon in combustion chamber	Remove cylinder head or cylinder and clean carbon from head and piston.
Loose or worn connecting rod	Replace connecting rod.
Loose flywheel	Check flywheel key and keyway; replace parts if necessary. Tighten flywheel nut to proper torque.
Worn cylinder	Replace cylinder.
Improper magneto timing	Time magneto.

ENGINE MISSES UNDER LOAD

Cause	Remedy
Spark plug fouled	Clean and regap spark plug.
Spark plug porcelain cracked	Replace spark plug.
Improper spark plug gap	Regap spark plug.
Pitted magneto breaker points	Clean and dress breaker points. Replace badly pitted breaker points.
Magneto breaker arm sluggish	Clean and lubricate breaker point arm.
Faulty condenser (except on Tecumseh Magneto)	Check condenser on a tester; replace if defective.
Improper carburetor adjustment	Adjust carburetor.
Improper valve clearance (four-stroke cycle engines)	Adjust valve clearance.
Weak valve spring (four-stroke cycle engines)	Replace valve spring.
Reed fouled or sluggish (two-stroke cycle engines)	Clean or replace reed.
Crankcase seal leak (two-stroke cycle engines)	Replace worn crankcase seals.

ENGINE LACKS POWER

Cause	Remedy
Choke partly closed	Open choke.
Improper carburetor adjustment	Adjust carburetor.
Magneto improperly timed	Time magneto.
Worn piston or rings	Replace piston or rings.
Lack of lubrication (four-stroke cycle engine)	Fill crankcase to proper level.
Air cleaner fouled	Clean air cleaner.
Valves leaking (four-stroke cycle engine)	Grind valves.
Reed fouled or sluggish (two-stroke cycle engine)	Clean or replace reed.
Improper amount of oil in fuel mixture (two-stroke cycle engine)	Drain tank; fill with correct mixture.
Crankcase seals leaking (two-stroke cycle engine)	Replace worn crankcase seals.

ENGINE OVERHEATS

Cause	Remedy
Engine improperly timed	Time engine.
Carburetor improperly adjusted	Adjust carburetor.
Air flow obstructed	Remove any obstructions from air passages in shrouds.
Cooling fins clogged	Clean cooling fins.
Excessive load on engine	Check operation of associated equipment. Reduce excessive load.
Carbon in combustion chamber	Remove cylinder head or cylinder and clean carbon from head and piston.
Lack of lubrication	Fill crankcase to proper level.
Improper amount of oil in fuel mixture	Drain tank; fill with correct mixture.

ENGINE SURGES OR RUNS UNEVENLY

Cause	Remedy
Fuel tank cap vent hole clogged	Open vent hole.
Governor parts sticking or binding	Clean, and if necessary, repair governor parts.
Carburetor throttle linkage, throttle shaft and/or butterfly binding or sticking	Clean, lubricate, or adjust linkage and deburr throttle shaft or butterfly.

ENGINE VIBRATES EXCESSIVELY

Cause	Remedy
Engine not securely mounted	Tighten loose mounting bolts.
Bent crankshaft	Replace crankshaft.
Associated equipment out of balance	Check associated equipment.

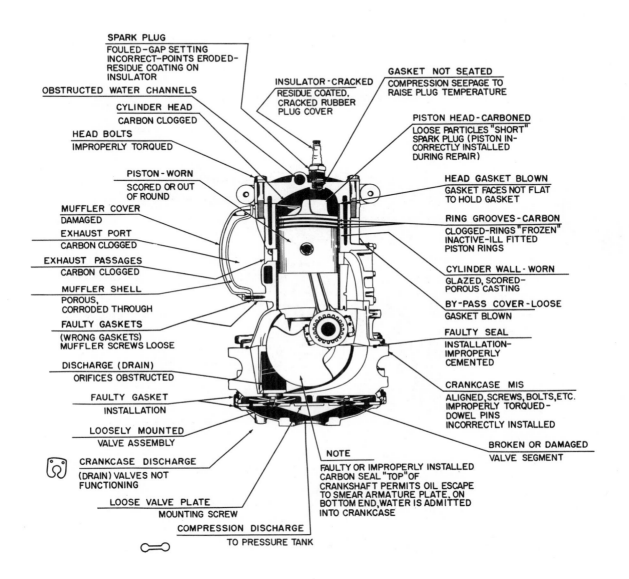

SPARK PLUG
FOULED-GAP SETTING
INCORRECT-POINTS ERODED-
RESIDUE COATING ON
INSULATOR

OBSTRUCTED WATER CHANNELS

CYLINDER HEAD
CARBON CLOGGED

HEAD BOLTS
IMPROPERLY TORQUED

PISTON - WORN
SCORED OR OUT
OF ROUND

MUFFLER COVER
DAMAGED

EXHAUST PORT
CARBON CLOGGED

EXHAUST PASSAGES
CARBON CLOGGED

MUFFLER SHELL
POROUS,
CORRODED THROUGH

FAULTY GASKETS
(WRONG GASKETS)
MUFFLER SCREWS LOOSE

DISCHARGE (DRAIN)
ORIFICES OBSTRUCTED

FAULTY GASKET
INSTALLATION

LOOSELY MOUNTED
VALVE ASSEMBLY

CRANKCASE DISCHARGE
(DRAIN) VALVES NOT
FUNCTIONING

LOOSE VALVE PLATE
MOUNTING SCREW

COMPRESSION DISCHARGE
TO PRESSURE TANK

INSULATOR - CRACKED
RESIDUE COATED,
CRACKED RUBBER
PLUG COVER

GASKET NOT SEATED
COMPRESSION SEEPAGE TO
RAISE PLUG TEMPERATURE

PISTON HEAD - CARBONED
LOOSE PARTICLES "SHORT"
SPARK PLUG (PISTON IN-
CORRECTLY INSTALLED
DURING REPAIR)

HEAD GASKET BLOWN
GASKET FACES NOT FLAT
TO HOLD GASKET

RING GROOVES - CARBON
CLOGGED-RINGS "FROZEN"
INACTIVE-ILL FITTED
PISTON RINGS

CYLINDER WALL - WORN
GLAZED, SCORED-
POROUS CASTING

BY-PASS COVER - LOOSE
GASKET BLOWN

FAULTY SEAL
INSTALLATION-
IMPROPERLY
CEMENTED

CRANKCASE MIS
ALIGNED, SCREWS, BOLTS, ETC.
IMPROPERLY TORQUED-
DOWEL PINS
INCORRECTLY INSTALLED

BROKEN OR DAMAGED
VALVE SEGMENT

NOTE
FAULTY OR IMPROPERLY INSTALLED
CARBON SEAL "TOP" OF
CRANKSHAFT PERMITS OIL ESCAPE
TO SMEAR ARMATURE PLATE, ON
BOTTOM END, WATER IS ADMITTED
INTO CRANKCASE

Fig. A-1 Power Head Diagnosis Chart.

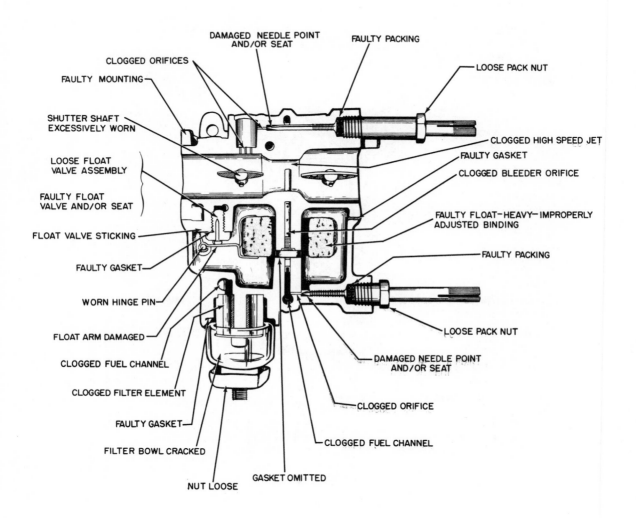

Fig. A-2 Carburetion Diagnosis Chart.

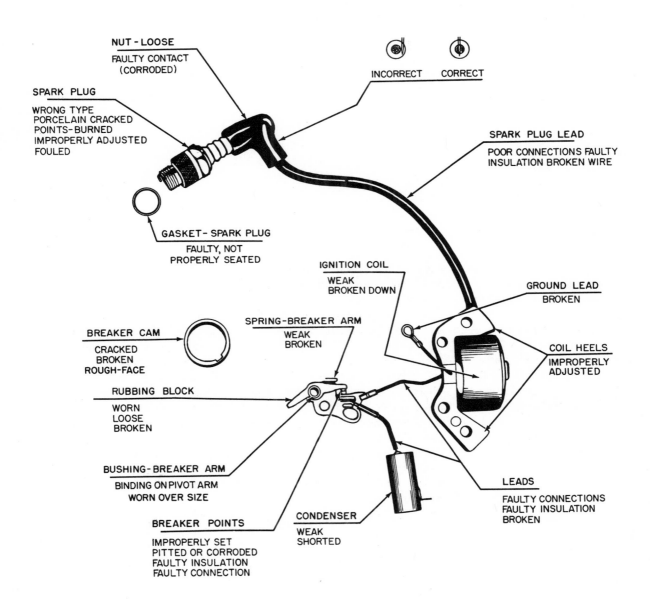

Fig. A-3 Magneto Diagnosis Chart.

Acknowledgments

Publications Director
    Alan N. Knofla

Editor-in-Chief
    Marjorie Bruce

Sponsoring Editor
    Frederick W. Smith

Revision
    George E. Stephenson

Production Director
    Frederick Sharer

Production Specialists
    Lee St. Onge
    Jean Le Morta

Illustrators
    Anthony Canabush
    Michael Kokernak

Many manufacturers and government agencies have supplied information in the form of instruction books, books on basic principles, pamphlets, booklets, and photographs. Without these sources of firsthand information, the research and writing of this book would have been difficult, if not impossible.

The author wishes to acknowledge the following companies and organizations which have supplied material and information: Aerospace Industries Association of America, Inc., Automotive Electric Association, Bolens Products Division Food Machinery and Chemical Corp., Buick Motor Division General Motors Corp., Bellows-Valair, Cedar Rapids Engineering Co., Compressed Air and Gas Institute, Cushman Motor Works, Inc., Delco-Remy Division General Motors, E. Edelmann and Co., Fluid Power Society, Forester Brothers, Gravely Tractors, Inc., Holley Carburetor Co., International Rectifier, Internal Combustion Engine Institute, Lawn Mower Institute, Inc., Marvel-Schebler Products Division Borg Warner Corp., Mustang Motor Products Corp., Northrop Corp., Oliver Outboard Motors, Outboard Motor Manufacturer's Association, Propulsion Engine Corp., Quick Manufacturing, Inc., Racine Hydraulics and Machinery, Inc., The Garrett Corp., Thomson Products, Toro Manufacturing Corp., West Bend Aluminum Co., Whizzer Industries, Inc., Wright Saw Division Thomas Industries, Inc.

Particular thanks is due to those companies and government agencies which have supplied many of the illustrations that are found in the text. Specific contributions are listed below.

AC Spark Plug Division, General Motors Corporation, Flint, Michigan — figures 7-25, 7-26

Aero Jet General Corporation — figures 24-2, 24-8

Aluminum Company of America, Pittsburgh, Pa. — figure 3-7

American Oil Company — figure 4-17

Atomic Energy Commission, Oak Ridge, Tennessee — figures 22-1 through 22-12

Automotive Manufacturers Association, Inc. — figures 12-24, 25-2, 25-3, 25-4

Bell Telephone Laboratories, New York, New York — figures 23-1 through 23-8

Briggs - Stratton, Milwaukee, Wisconsin — figures 3-5, 3-17, 3-24, 3-25, 4-3, 4-4, 4-5, 4-9, 4-10, 4-24, 4-28, 5-7, 7-9, 7-19, 8-7, 8-8, 8-9, 9-13, 9-20, 9-21, 9-22, 9-28, 9-30, 9-31

Champion Spark Plug, Toledo, Ohio — figures 4-21, 4-22, 4-23, 7-24, 8-2, 8-3, 8-4, 8-5

Chevrolet Division, General Motors Corporation, Detroit, Michigan — figures 12-5, 12-6, 12-8, 12-9, 12-27, 12-30

Clinton Engines, Clinton, Michigan — figures 3-6, 3-8, 3-14, 3-15, 3-16, 3-27, 4-1, 4-2, 5-4, 8-12, 8-13, 8-14, 8-15, 8-16, 8-17, 9-1, 9-2, 9-5, 9-6, 9-7, 9-8, 9-9, 9-10, 9-11, 9-12, 9-23, 9-24, 9-25, 9-27, 9-34

Delco-Remy Division, General Motors Corporation, Anderson, Indiana — figures 12-18 through 12-21

Detroit Diesel Engine Division, General Motors Corporation, Detroit, Michigan — figures 13-3, 13-4, 13-5, 13-7, 13-8

Esso Research and Engineering Company, Linden, New Jersey — figures 23-12, 23-18

Ethyl Corporation — figure 4-20

Evinrude, Milwaukee, Wisconsin — figure 6-10

Fairbanks-Morse, Chicago, Illinois — figures 7-8, 7-10, 7-11, 7-13, 7-14, 7-20, 7-22

Ford Division, Ford Motor Company, Dearborn, Michigan — figures 12-1, 12-2, 12-3, 12-4, 12-7, 12-10, 12-11, 12-12, 12-13, 12-14, 12-15, 12-16, 12-17, 12-22

General Electric Company, Schenectady, New York — figures 14-3, 14-6, 14-15, 18-12, 23-1B

General Motors Corporation, Detroit, Michigan — figures 4-8, 16-12, 16-13, 16-14, 16-15, 18-1, 18-2, 18-3

International Telephone and Telegraph Corp., New York, New York —figure 15-1

Johnson Motors, Waukegan, Illinois — figures 3-3, 3-13, 4-6, 4-7, 4-13, 4-19, 5-13, 6-7, 6-8, 6-9, 7-12, 7-21, Appendix A-1, A-2, A-3

Kiekhafer Corporation, Fond du Lac, Wisconsin — figures 3-16, 9-17, 9-26, 9-32, 9-33, 9-35, 9-36, 9-37

Kohler Company, Kohler, Wisconsin — figure 3-15

Lauson Power Products Department, Division of Tecumseh Products, Grafton, Wisconsin — figures 4-26, 4-27, 5-6, 5-10, 5-11

Lawn-Boy Company, Galesburg, Illinois — figures 3-4, 6-2, 6-4, 9-4

Lincoln-Mercury Division, Ford Motor Company, Dearborn, Michigan — figures 3-1, 6-5, 6-6, 8-6

McCullock Company, Los Angeles, California — figure 4-12

McDonnell-Douglas, St. Louis, Missouri — figures 24-3, 24-4, 24-5, 24-6

National Aeronautics and Space Administration — figures 15-2, 15-3, 15-4, 15-6, 25-1A

Nuclear Division, Martin Company, Baltimore, Maryland — figures 23-1A, 23-1C, 23-1D, 23-2, 23-10

Perfect Circle Company, Hagerstown, Indiana — figures 3-9, 3-10, 3-11, 9-29

Pratt-Whitney Aircraft Company, Hartford, Connectcut — figures 14-2, 14-8, 14-9, 14-12, 14-14

Rocketdyne Division, North American Aviation, Canoga Park, California — figures 15-5, 15-11

Standard Oil Company, of New Jersey, New York, New York — figures 3-19, 3-20, 3-21, 3-22, 4-19, 5-5, 6-1

Tecumseh Products Company, Grafton, Wisconsin — figures 3-12, 3-23, 3-26, 3-28, 3-29, 4-16, 4-22, 7-23, 8-10, 8-11, 9-14, 9-15, 9-16, 9-19

Terry Steam Turbine Company, Hartford, Connecticut — figures 18-8, 18-9, 18-10

United States Air Force, Washington, D.C. — figures 14-1, 14-4, 14-5, 14-7, 14-10, 14-11, 14-13, 15-8, 15-9, 15-10, 15-12

Westinghouse Electric Company, Philadelphia, Pennsylvania — figure 24-7

Wico Electric Company, West Springfield, Massachusetts — figures 7-16, 7-17, 7-18

Wisconsin Motor Corporation, Milwaukee, Wisconsin — figures 4-25, 5-8, 5-9, 6-3

John Wiley Company (Thermal Engineering), New York, New York — figures 18-5, 18-6, 18-7

Zenith Carburetor Division, Bendix Corporation, Detroit, Michigan — figures 4-14, 4-15

INDEX

INDEX